Print is Dead

Print is Dead

Books in Our Digital Age

Jeff Gomez

palgrave
macmillan

PRINT IS DEAD
Copyright © Jeff Gomez, 2008.

All rights reserved.

First published in hardcover in 2008 by PALGRAVE MACMILLAN® in the US - a division of St. Martin's Press LLC, 175 Fifth Avenue, New York, NY 10010.

Where this book is distributed in the UK, Europe and the rest of the world, this is by Palgrave Macmillan, a division of Macmillan Publishers Limited, registered in England, company number 785998, of Houndmills, Basingstoke, Hampshire RG21 6X

Palgrave Macmillan is the global academic imprint of the above companies and has companies and representatives throughout the world.

Palgrave® and Macmillan® are registered trademarks in the United States, the United Kingdom, Europe and other countries.

ISBN: 978–0–230–61446–8

A catalog record for this book is available from the Library of Congress.

A catalogue record for this book is available from the British Library.

First PALGRAVE MACMILLAN paperback edition: June 2009

10   9   8   7   6   5   4   3   2   1

Printed and bound in the United States of America.

contents

contents

For my loving wife, Colclough, who heard most of these ideas as soon as they popped into my head... and never once told me to shut up

One of the questions that haunts me – it's a question for philosophers and brain science – is, if you've forgotten a book, is that the same as never having read it?

Tom Stoppard

introduction

IN THE 1984 FILM *Ghostbusters*, Annie Potts (a secretary at the newly opened Ghostbusters office in Manhattan) asks nerdy scientist Harold Ramis whether or not he likes to read. Ramis doesn't hesitate for a second; to him, the answer is so obvious it's not even a question. His reply? 'Print is dead.' When the movie was first in theaters, well over twenty years ago, people laughed at these words. Since Ramis a few seconds later acknowledges his hobbies to include collecting molds, spores and fungus, his statement about print being 'dead' was meant to be equally nonsensical and outrageous. Right? Because print can't possibly be dead; it's everywhere.

After all, books and newspapers exist in abundance and people read them avidly each day. Mounds of printed material sit in stacks upon stacks in stores, offices and homes across the country and the world. You rarely see someone taking a form of mass transportation – bus, commuter train, etc. – who isn't clutching some sort of reading material (indeed, in New York City people look forward to riding the subway so they can catch up on their reading; and if they don't have something to read, they just read over the shoulder of the person next to them). The same goes for the hundreds of thousands who travel each day by plane, flying for either business or pleasure. Most airports have a bookstore or newsstand, if not both. And then there are the millions of newspapers that are delivered every morning to doorsteps all over the world, folded into thirds like a wallet of information just waiting to be opened. There are even bestseller lists that appear every week in numerous newspapers as a kind of year-round Olympic competition, proudly stating which books we're all buying instead of all the other books we're not buying.

Whole rooms of houses and apartments are lined with books, and some homes even have their own libraries stacked floor-to-ceiling with row after row of hardbacks and paperbacks. Furniture and office supply stores feature a dizzying array of bookshelves, while corner kiosks in large cities are like ice cream trucks for printed material, selling glossy magazines and black-and-white newspapers. So with all of this ink-on-paper floating around, read day after day and being so ingrained and intertwined in all of our lives – not to mention that the publishing industry is a billion-dollar a year business – print *can't* be dead, can it?

While print is not yet dead, it is undoubtedly sickening. News-paper readership has been in decline for years, magazines are also in trouble, and trade publishing (the selling of novels and non-fiction books to adults primarily for entertainment), has not seen any substantial growth for years. More and more people are turning away from traditional methods of reading, turning instead to their computers and the Internet for infor-mation and entertainment. Whether this comes in the form of getting news online, reading a blog, or contributing to a wiki, the general population is shifting away from print consump-tion, heading instead to increasingly digital lives.

'In less than half a century we have moved from a condition of essential isolation into one of intense and almost unbroken mediation,' wrote Sven Birkerts in the updated 2006 edition of his book *The Gutenberg Elegies: The Fate of Reading in an Electronic Age*. 'A finely filamented electronic scrim has slipped between ourselves and the so-called "outside world." The idea of spending a day, never mind a week, out of the range of all our devices sounds bold, even risky.'

Because of all these devices, people are more connected than ever. Whether it's through Blackberries, cell phones, or smaller and more portable laptop computers – aided by the high-speed connections most people have at the office and at home – hundreds of millions of people around the globe are electronically linked, and the concept of ever being 'offline' is becoming increasingly rare.

A society that used to communicate via the postal service, expecting someone to respond to a letter in a number of days, now expects a response to an email in a matter of seconds. And in terms of their news and information, a society that was once content with the evening news and the morning news-paper now demands access to events every second of the day. This new way of living has had both positive and negative repercussions for most of the entertainment industries, nota-bly music, television and film. Finally, these seismic digital shifts are being felt by publishers. It is now impossible for the world of books not to be affected by the immense cultural changes that a digital world has set into motion. And if writ-ers, agents and publishers don't adapt their business models and practices to embrace and accept these changes, they stand to lose both their existing audience and their steady rev-enue. The publishing industry, which hasn't faced a significant challenge in decades, now has to do the thing it likes least: change.

Meanwhile, the needs of an entire generation of 'Digital Natives' – kids who have grown up with the Internet, and are accustomed to the entire world being only a mouseclick away – are going unanswered by traditional print media like books, magazines and newspapers. For this generation – which Googles rather than going to the library – print seems expen-sive, a bore and a waste of time. What can a book give them that a blog or website can't? What's the point in going to a Barnes & Noble, or why should they have to wait until their shipment arrives from Amazon? Why not just go to iTunes and download half a dozen new songs, or send a friend a video shot on their cell phone instead? These are questions that the publishing industry must face.

Of course, neither the 'print is dead' argument nor the 'future of the book' debate are new topics. Both discussions have been floating around for decades, gaining critical mass first in the mid-1980s when personal computers became commonplace, and then again in the 1990s at the dawn of the Internet. Back then, these were primarily abstract debates. But the developed world's nearly totally digital society now necessitates putting the debate at the forefront of public discussion. And while many forms of art have been declared 'dead' over the years (the death of the novel was discussed as early as the 1960s), what was being debated in those instances was whether or not a certain school of painting or style of writing had anything new or relevant to say. No one was arguing that books or canvasses themselves – whatever they contained – would be the things to disappear. But in the past few years there have been new ideas about how printed words and books themselves will have to change and adapt to a new digital reality, and this book is an attempt to sort out and filter those various discussions.

In my own case, I know in my heart that I am who I am because of books, because of the words of others that I discovered between the hardback and paperback covers of worn and dog-eared novels. Whole worlds unfurled at the touch of my fingertips when I read good books. I met characters and experienced things that informed how I thought and felt about everything: life, love, death etc. I later became a writer, and was fortunate enough to have several novels published (long ago) by a major publishing company. It was the thrill of a lifetime to see my books on shelves, or my name quoted in reviews that appeared in newspapers I'd been reading all my life. But when the writing career didn't really work out I got into the business side of things, moving to New York and getting a job in publishing. I've now spent the last ten years thinking about words instead of stringing them together into fictional sentences.

In 1999, when the industry was just getting off the ground, I worked in eBooks and electronic content. A few years later, during the rise and eventual dominance of the Web, I segued into online marketing. During all of this, witnessing how the Internet was changing nearly every aspect of our daily lives, I began to think – and tried to predict – what these changes would mean to the world of writers, readers and publishers. I have now come full circle, and find myself writing a non-fiction book about, well, *books*.

In terms of *Print is Dead*, what I've tried to do is lay out my argument in a concise, three-part structure that first shows how publishing needs to change. I then describe the current conditions in terms of what's happening in other industries and what brought us culturally to this point. Finally, I discuss the issues going forward in terms of what life will be like in a digital world for writers, readers and publishers.

I outline all this to set expectations, since there's a lot that pertains to the future of the book debate that's *not* in this book. For instance, I don't really go into what digital reading will mean to libraries or universities, or to different kinds of publishing, such as academic, educational or children's. I also don't get into the specific economics of book publishing, nor do I wade into the debate over Google's program to scan books and libraries. I avoided these areas since I felt that to weigh in on each of them would only bloat the book and thus diffuse the argument. I wanted *Print is Dead* to be lean in its physical form and concise in its scope. Also, the issues swirling around Google are time-sensitive, and are more suited for a magazine article than a book (indeed, there have already been many such articles), while the business and economics of publishing are more suited to a textbook.

Also, while I realize that the debate over the future of the book is a global discussion (wherever in the world you find readers, they will be touched by this subject), mine is a New York-centered book. This is because I'm a New York-centered guy, and this is what I know. I work in publishing in Manhattan, and *The New York Times* is the paper I read (online) every morning. And while I can see lots of value in a book that covers the global print scene, writing up dispatches from the farthest reaches of our recently flattened world, I'm not the person to write that book. That's more of a Tom Friedman 'I'm here in Bangalore after visiting China' kind of thing; my passport just doesn't have all of his stamps.

So rather than being an exhaustive and comprehensive look at every nook and cranny of contemporary publishing, this book is meant to be more a catalyst for change and an encapsulation of the overall argument, a tidy one-stop shopping experience for the future of the book debate. I happily let the specialists and scholars fill hundreds of pages on things like copyright law or balance sheets.

Finally, if you're either chuckling or frowning as you read all this, thinking that print can't be *that* dead if – here you are – reading words on paper you can hold in your hands, this fact is not lost on either me or my publishers. It reminds me of an episode of *The Simpsons* from the mid-1990s in which the recurring character Sideshow Bob, upset at the cultural destruction caused by television, threatens to blow up Springfield if TV is not immediately taken off the air. To issue his threat, Sideshow Bob appears on a large television screen during an air show, warning the townspeople and its elected officials that he'll destroy the entire population if his demands aren't met. Right after signing off, Bob reappears and grouchily touches upon the absurdity of the situation, telling the crowd, 'By the way, I am aware of the irony of appearing on television in order to decry it. So don't bother pointing that out.'

stop the presses

1

byte flight

A BLANK PIECE of paper and a computer screen when it's off have something in common: both are empty, devoid of content, ripe with possibility. A myriad of things could cover each: words, numbers, pictures; philosophy, comedy, tragedy. The possibilities are endless.

But while you can only fit so much onto one piece of paper (only so many words and so many numbers, no matter how small you write), a computer screen can be an inexhaustible source of endless information. A computer screen is a gateway, forever replenishing itself by either scrolling or replacing old information with new. A piece of paper is merely a clean slate. We can also see our reflections in screens, while paper is always opaque; no matter how long you stare into a blank piece of paper, you'll never see yourself.

Screens, and the possibilities they represent, now dominate our lives more than ever before. From large ones in the form of high-definition LCD televisions that explode to the size of a small movie screen (with a theater sound system to match) to the tiniest of screens on our cell phones that keep us in contact with the world through texting, ringtones and even videos. And then there are the screens of computer monitors, from desktop machines sitting in offices to notebooks that fit neatly into backpacks and are transported everywhere.

Many cities have gone completely wireless, while in some towns a number of businesses offer free wireless connectivity. Most apartment buildings in big cities are hot spots, with half the tenants sporting a wireless connection (while the other half surreptitiously use the signal). Yet while screens offer us unlimited potential and limitless boundaries, with people using them every day of their lives to deliver and consume information, there is one area of knowledge that is sacrosanct, seen to be both untouchable in terms of its utility and unimpeachable by its very nature. I'm talking about the words found in books.

People love books. They love the way books look on their shelves, coffee tables and nightstands. People love the way books feel in their hands, and they even love the way books smell (well, the old ones yellowed with age and cured with dust). Booklovers the world over have spent countless hours caressing covers, fingering dust-jackets, and repeatedly running their hands up and down fabric spines. Homes everywhere house collections of the hardback and paperback covers of worn and dog-eared novels that bibliophiles have been amassing for years, moving them in near-bursting boxes from place to place over the span of their adult lives.

Because of this level of devotion, and the fact that the people who love books love them in the way that patriotic people feel about flags or musicians feel about their instruments, how can books ever be replaced, let alone disappear? And is a computer screen really going to substitute for books in the hearts or minds of, well, *anybody*?

The answer to this would seem to be an easy *no*, and the idea of books becoming more and more rare until they're finally erased from existence sounds ludicrous to most people. Yet this shift is already subtly happening all around us, and books are indeed on the way out, while screens keep inching their way in.

'In many ways, we have, for better or worse, already moved beyond the book,' wrote George P. Landow in the 1996 essay collection *The Future of the Book*. 'Even on the crudest, most materialist standard involving financial returns, we no longer find it at the center of our culture as the primary means of recording and disseminating information and entertainment. The sales of books and other printed matter, for centuries the center of our technology of cultural memory, now have fallen to fourth position behind the sales of television, cinema, and video games.'

In this new, increasingly digital age – where the limits of communication and information have been shattered – the idea of isolating words between the covers of a book seems not only quaint but anachronistic. Pages are cages, trapping words within boundaries. And while booklovers may still love their books, choosing to ignore what computers and screens could mean to the world of words, a new generation is already bypassing print for electronic alternatives, getting all or most of its information from the Web.

To many people, for whom books have played a crucial part in their lives and in the development of their personalities, this is difficult to accept. For most writers today over the age of thirty, it was the discovery of books – and their own hunt for interesting-looking volumes in bookstores – that made them want to be writers in the first place. Yet, for that new generation who gets all of its information over the Internet, inspiration is coming from elsewhere. Personalities, careers and works of art are now being downloaded instead.

And it's no longer just the early adopters and the techno-geeks fueling the 'print is dead' debate. Instead, flat sales and growing trends are solidifying into everyday behavior and new business models. Print is beginning to look seriously endangered, and there are those who speak of it eventually going the way of the dodo.

New technologies and practices are already disrupting the carefully cultivated status quo of numerous other media and entertainment industries, such as movies, newspapers, magazines and television; the repercussions for print are plain to see. Consider the following statistics for the year 2005, taken from Chris Anderson's bestselling book *The Long Tail: Why the Future of Business Is Selling Less of More*:

- Hollywood box office fell 7 per cent, continuing a decline in attendance that started in 2001 and appears to be accelerating.
- Newspaper readership, which peaked in 1987, fell by 3 per cent (its largest single-year drop) and is now at levels not seen since the 1960s.
- Magazine newsstand sales are at their lowest level since statistics have been kept, a period of more than thirty years.
- Network TV ratings continue to fall as viewers scatter to cable channels; since 1985, the networks' share of the TV audience has fallen from three-quarters to less than half.

This is the future that trade book publishing has to look forward to: declining sales and decreased market share. And while the shrinking audiences for the major entertainment outlets are beginning to reappear online (consumers either renting DVDs or downloading movies, reading newspapers and magazines online, and streaming TV shows from the networks' websites or file sharing sites), the loss of interest in reading and literary culture is not being replaced. Instead, the printed word is being supplanted by any number of other online attractions: YouTube, MySpace, role-playing videogames, and of course all of those millions and millions of iPods. In fact, it is iPods – and what they've done to the music business – that may offer the best glimpse of the future of the book.

To see where words are headed, simply follow the evolution of music's various technological leaps from one format to another: wax cylinder, vinyl, eight track, cassette, compact disc, MiniDisc, MP3. What's important to note in this sequence is that the last format – MP3 – doesn't necessarily exist. It's a file format, a way of digitizing and storing information. It's not a physical thing that you necessarily hold or trade. MP3s do not collect dust and will not be sold years from now at a flea market. The majority of printed material will eventually undergo a similar transformation, ending up as a digital file instead of a physical thing.

Since the 1500s the evolution of books has become stuck at one of the early stages (somewhere between vinyl and eight track). Books have evolved from scrolls to hand-illuminated calf-skin manuscripts to bound paper volumes, but except for minor variations – such as trade and mass market paperbacks – there has been very little change in books (or in the book business) in hundreds of years. While new ways of sharing stories have proliferated – from stream-of-consciousness prose to the recent rise of graphic novels – there has been a shortage of innovation when it comes to selling stories.

This continued ubiquity of the printed book has lulled readers and publishers into a false sense of security. Because books have been this way for *so* long, the industry is unable to envision anything different from what currently exists. Yet it's clear to see that the evolutionary stage for books will be much the same as for music: its final format will not physically exist. Instead, the majority of printed words will evolve into electronic files that will be distributed, bought, sold, and consumed on digital devices.

While this idea may scare certain authors, many of whom have grown to think of their work solely in terms of the physical artifacts created by publishers, many more of them won't ultimately care. These authors will learn from the musicians who have seen their music travel from LPs through CDs, finally landing in the ether of MP3s; whatever the format, they know that the song remains the same. Just as Pink Floyd's *Dark Side of the Moon* no longer means a black vinyl disc inside a cardboard sleeve with a prism on the front, but is available instead as digital download through iTunes, J. D. Salinger's *The Catcher in the Rye* will one day no longer be associated with a stack of cream paper and a plain, paperback cover. Music divorced from packaging is all about the songs, not the disc it comes on or the sleeve it comes in; novels and other works of literature will be all about their words.

Critics who can't believe this, that great novels will still be great novels when separated from their physical packages, are much like The Grinch in Dr Seuss's classic story *The Grinch Who Stole Christmas*. The Grinch looked at Christmas and saw only the surface: wrapping paper and colorful lights, piles of toys and yards of decoration. And yet – as he discovered after he stole all of the trappings of the holiday from the townspeople of Whoville – Christmas continued to exist, perhaps more at that point than ever. 'It came without ribbons! It came without tags!,' the Grinch finally discovered. 'It came without packages, boxes or bags!' The same could be said for books that come without pages or covers; their greatness will continue to shine, and they will still be great works of literature.

In the introduction to his 1984 book, *99 Novels* – which was Anthony Burgess's pick of the best novels released between the start of the Second World War and the year we were all supposed to wake up being watched by Big Brother – the prolific English writer grappled with the question, 'What is a novel?' He finally gave up, realizing that, 'the only possible answer is a shrug.' Realizing that books are primarily a commodity, he finally decides, 'the practical answer is provided by the publishers, printers and binders who process a manuscript into a printed copy dressed in an overcoat.' He continues:

But at this moment in history I have to accept, with everyone else, that a novel is a visual experience – black marks on a white page, many of these bound into a thickish book with a stiff cloth cover and an illustrative dust-jacket. Its paperback version is a poor but necessary thing, a concession to the pocket, the sickly child of the original. When we think of *War and Peace* or *David Copperfield* we see a fat spine with gold lettering, the guardian of a great potentiality (signs turned into sense), proudly upright on a shelf. BOOK can be taken as an acronym standing for Box of Organized Knowledge. The book called a novel is a box from which characters and events are waiting to emerge at the raising of the lid.

Burgess was on the right track in realizing that a physical book is merely a container, and that its printed form and shape is a concession to the marketplace. (He also – playing the futurist – wrote, 'soon we may get our novels on floppy disc.') If this is the case, then the *box* portion of BOOK's *box of organized knowledge* is really just a container. What's important is the *knowledge*, and most of this knowledge can be contained in a variety of digital formats that are much more efficient than a simple 'box' of physical print. These electronic formats, as well as amazing new levels of interactivity, are just beginning to be created and widely experienced, but their arrival will be the book's biggest leap of evolution since Gutenberg. And it won't just be fiction and non-fiction; there are numerous other areas of publishing changing in the wake of our impending digital future: education and textbook publishing, newspapers and magazines, not to mention public libraries and college campuses. Digitization will free all forms of these previously contained (and constrained) stores of knowledge from their former boxes, where they will be consumed and enjoyed in ways that Burgess (not to mention Gutenberg) couldn't possibly have imagined.

As part of the 2007 London Book Fair there was a panel entitled 'Digitise or Die: What is the Future of the Author?' Among the panelists was noted novelist Margaret Atwood. Here's a description of the seminar, taken from the London Book Fair's website: 'New technology is finally outstripping the revolution caused by the printing press. What is the future of the book? How do publishers take advantage of the new technology and protect the author's interests at the same time? Are publishers dragging their heels in the face of the inevitable? How should writers respond to it? What are the new opportunities for writers in the electronic revolution? And how does all this affect the reader?'

During the seminar, it seems that Atwood made the usual points, such as eBooks will be good for reference material but not much else. She also criticizes the utility of eBooks, saying that they can't be read in the bathtub. (Actually, I'd say that printed books – unless you want them ruined – are also pretty difficult to read in a bathtub.) Of course, the dozens of things that eBooks *can* do, that print books cannot, didn't get mentioned. Instead, the argument boiled down to the usual emotional tug of the object itself, with panelist Philip Zimbardo declaring about a book: 'It's something you hold, near to your heart.'

Actually, you hold a book near to your eyes, so it can be read. If you hold it near to your heart what you're then doing is giving it a hug (which technically doesn't count as reading). When you give a book a hug, or show it love, you're no more showing love to that physical thing than you are to a photograph when you kiss a picture of a sweetheart. What you're reacting to, when gazing at a snapshot, is the subject of the picture and not the photograph itself. The same goes for books; it's a writer's words that touch us, not the paper those words were printed on.

The idea that computers are cold, impersonal things that we'll never be able to interact with, comes up often in the discussions of the book's future. Most bibliophiles run in the opposite direction anytime the words 'digital' and 'book' are used in the same sentence. More and more, what I call 'byte flight' is becoming part of the debate over the future of reading, with computers cast as the perennial bad guys while books are the object that can – by their very nature – do no wrong.

Technology writer Mike Elgan, writing on the website for *Computerworld* magazine, had an article in 2007 not long after the London Book Fair unambiguously titled 'Why e-books are bound to fail' (if the main headline left you wondering about Elgan's thoughts on the issue, the sub-headline should remove all doubt: 'Electronic books pack bleeding-edge technology, too bad they'll never catch on'). Elgan, in going through his reasons for why he thinks eBooks are dead in the water, touches upon all of the usual reasons: price, format and device confusion etc. But in the end, Elgan's big show-stopping pronouncement is that 'people love paper books.' And because of this, 'e-books are not, and cannot be, superior to what they are designed to replace.'

Yes, some people love paper books. But other people have never thought about a book in their life, and simply want the stories and ideas that books contain. I doubt your average reader of Danielle Steel or James Patterson collects rare books. Rather, they collect the emotions that they get from Danielle Steel and James Patterson novels, and this is what keeps them buying book after book. So to talk about one format being superior to another is silly; this isn't a joust, it's about utility, and the fact is that electronic books can do plenty of things that paper books cannot do.

Beyond this, I really do think it's wrong for a technology writer (not to mention one writing for a publication entitled *Computerworld*) to write a sentence such as 'Unfortunately, these [eBook] products – as well as the whole product category – are destined for failure.' Elgan has little to back up his claim beyond the usual facile arguments, including the fact that people will never want to '"curl up" with a battery-operated plastic screen.' I don't know what kind of computers populate Elgan's world, but people 'curl up' with 'battery-operated plastic screens' all the time. What does he think an iPod or a laptop is? Or a Blackberry? These items are now an entire generation's prized possessions. Most kids today 'curl up' with nothing but battery-operated plastic screens (when they're not curling up with other teens, that is). For them, these aren't even gadgets, but instead are everyday objects that form an essential part of their young lives. The same goes presumably for *Computerworld*'s subscribers and readers. So no matter how much people 'love paper books,' we indeed live in a computer-filled world, leading primarily digital lives, and the world of literature will eventually yield to 'battery-operated screens' the same way that music did.

In fact, Andrew Marr, writing on the *Guardian* website in 2007, had an essay entitled 'Curling up with a good ebook.' The premise of the essay was that Marr, 'who treasures his smelly, beautiful library of real books,' would spend a month reading the Sony eReader, a dedicated reading device that debuted in 2006. From the outset Marr is – like lots of readers – skeptical: 'If you are selling ebooks, I'm a hard sell. For one thing, my enthusiasm for traditional books is just this side of pervy. I live among mountains of them and always have, among the most beautiful mass-produced objects of all time.' And so even though Marr is an unabashed lover of books, even from a technical point of view ('books are such good technology, even compared with CDs or newspapers'), even he manages to see utility in electronic reading, as well as books existing as invisible computer files rather than physical things: 'In our house, every day we get mounds of newsprint, much of it thrown instantly away. The stuff hangs around like intellectual scurf, and it's depressing.'

He starts by taking the device to a number of locations, 'reading some Tolstoy and then some Conan Doyle, in the garden, slumped in a chair inside, on a sofa in a dimmish room, and in the back of a car.' To his surprise, reading text on a screen is not as bad as he thinks: 'In each place, it was easy to read; I have spent plenty of time reading it and so far, haven't felt any eyestrain, or no more than I would have found with a book.'

In the story there was even an accompanying photo that showed Marr curling up quite comfortably on a sofa, near a window, reading his eBook in a manner that most critics say is impossible. After a few weeks, Marr is pleased by the experience: 'I was surprised by how easy it was to use, and surprised by how much can be stored on it. I liked the rather elegant, retro design, more like a digital slate than a piece of flashy gear...' And even though he still had some issues, notably price and the experience of turning 'pages,' in the end, Marr remarked, 'I am reluctantly impressed with my ebook.'

I think if more people actually *tried* to read a book electronically, they'd have the same reaction. Instead, eBooks are shunned in the same way that films that sound blasphemous are always picketed before they're actually seen.

John Lanchester, also writing in the *Guardian* that same year, had an interesting essay entitled 'It's a Steal,' which talked about the problem of establishing a worth for literary content on the Web. Setting the stage, Lanchester writes: 'The revolutionary impact of the internet on the music and film business is plain to see. Now it's the turn of the printed word. The question is simple, and far-reaching: what's going to happen to books and to the people who write them?' The essay is long, and makes many good points, encompassing everything from the origin of copyright law to the efforts by Google to digitize books.

In the end, however, Lanchester relies on the usual pro-book arguments, and here's where he – for me – strikes the wrong chord: 'Personally, I think that books are going to be OK, for one main reason: books are not only, or not primarily, the information they contain. A book is also an object, and a piece of technology; in fact, a book is an extraordinarily effective piece of technology, portable, durable, expensive to pirate but easy to use, not prone to losing all its data in crashes, and capable of taking an amazing variety of beautiful forms.'

What Lanchester cites as positives are in fact negatives. As a thing of beauty, in the opinion of most people, a Modern Library edition of *The Razor's Edge* will always win over an electronic reading device loaded with an eBook of *The Razor's Edge*. But when looked at in terms of technology, there's no comparison; even the most rudimentary electronic reading experience offers more features and overall utility than a print book does. So to make the argument that books are great technology (and don't crash and don't lose data, etc.) is the supreme kind of silliness, not to mention it becomes ultimately defensive in nature (because instead of saying what books will do, you end up trumpeting all the things they won't do). I also think that, in the scope of the discussion now occurring in the 'print is dead' debate, books are indeed primarily the information they contain. After all, isn't that what makes us choose one over the other?

Even those you would expect to believe in the future of reading on screen will sometimes surprise you that they still want – alongside all of their gadgets – a good old fashioned book. Science fiction writer and tech-guru Cory Doctorow, in a *Locus* magazine essay from 2007 entitled 'You *Do* Like Reading Off a Computer Screen,' talks about how, even though the future will bring the adoption of 'super-portable screens' which will be heavily used, 'most of us won't spend most of our time reading anything recognizable as a book on them.' Doctorow admits that this is obviously contradictory (after all, if everyone starts using Tablet PC-like devices to keep in touch, read websites and compose blogs, why wouldn't they also use them to read narrative fiction?), but he explains this away by defining the novel as something that is – by its very nature – digitally undigestible.

'The novel is an invention, one that was engendered by technological changes in information display, reproduction, and distribution,' writes Doctorow. 'The cognitive style of the novel is different from the cognitive style of the legend. The cognitive style of the computer is different from the cognitive style of the novel.'

I can see Doctorow's overall point, but the problem with his argument is that he's referring to the novel like it's a static, regulated thing; as if each novel were the same in content, tone and construction. And if he's talking about thick classics like *Tom Jones* or *Vanity Fair*, then I can see his point that a computer device is not contemplative enough for a meandering, slowly unfolding *bildungsroman*.

But what about shorter, experimental novels? Or else books that just beg to be jumped in and out of at different points, like Pessoa's *The Book of Disquiet*? This classic book (it's not really a novel) is constructed in short, pithy bursts; there's not much of a narrative, and each of its sections is self-contained. Reading it is not unlike reading a blog. Or imagine having a chapter a day of Orwell's *Down and Out in Paris and London* sent to your cell phone to be read on the train in the morning. There are thousands of similar works that would hardly suffer from being read on a screen. Quite the opposite: they're actually perfect for electronic reading.

Not to mention the fact that none of this takes into account the new kinds of fiction and literature that have yet to be written. Throughout his essay Doctorow keeps a running list of all of the other things he has been doing while writing ('In the ten minutes since I typed the first word in the paragraph above, I've checked my mail, deleted two spams, checked an image-sharing community I like,' etc.). His attention, like the attention of anyone who lives a wired, online life, has been shattered and now exists in a dozen places at once.

Gone are the days when a writer like Proust worked in a cork-lined room in order to keep the sound of the world away from his ears (and his concentration). Kids now reading Doctorow, who will become the next generation of writers, will have been raised on computers, blogs, MP3s, RSS, iPods, MySpace and YouTube. The works they create will have this electronic DNA woven throughout them. Because of this, these new worlds of fiction will most likely be perfectly suited to an electronic screen. In the same way that all authors must be men and women of their time, in our digital present a new generation of writers will create computer-influenced works whose words will belong on a screen in the same way that the words of Thackeray and Fielding once belonged on the page.

Shortly after Doctorow's piece appeared, *The Economist* ran an article entitled 'Not bound by anything,' which dealt with the future of books and asked the question, 'Now that books are being digitised, how will people read?' The digitization in question was Google's book program, but within the context of this *The Economist* managed to ask some important questions, including the very ones that form the foundation of the future of the book movement and the 'print is dead' argument: 'How, physically, will people read books in the future? Will technology "unbind" books, as it has unbundled other media, such as music albums? Will reading habits change as a result? What happens when books are interlinked? And what is a book anyway?'

The article then went on to talk about eBooks, the new paradigm of wikis, the popularity of iPods, and what all of this could mean for non-fiction (hyperlinks galore) not to mention all of the material (such as novellas) that have never really fit within the business of traditional publishing. Finally, however, after extolling all of the virtues of electronic books, the writer trots out the standard 'the book is perfect' argument: 'Most stories, however, will never find a better medium than the paper-bound novel. That is because readers immersed in a storyline want above all not to be interrupted, and all online media teem with distractions (even a hyperlink is an interruption).'

In the same way that Doctorow seemed to classify all novels as being the same, so did *The Economist* classify all readers as similar, noting that the most important thing to them was to not be interrupted while they're reading. This is a silly if not insane notion. Readers are changing just as much as novels are, and have been for generations.

When *Less Than Zero* first appeared, more than twenty years ago, it was noted that its brief scenes, short chapters, and streamlined prose made it a novel for the MTV generation whose attention had been shattered and for whom slow narratives had been banished in order to make way for the three-minute fix of rock videos. Its author was 21-year-old Bret Easton Ellis, still in college when he wrote and published the book; it was the perfect chronicle for a new generation, written by a member of that same generation.

A quarter of a century later society has only increased its speed; we have not slowed down. Music and films reflect this, and the rise of reality television has shown that audiences no longer have the patience for storylines, characters or plot. Today's readers (not to mention tomorrow's) are used to email, instant messaging; blogs, podcasts, and a dozen other inventions that didn't exist a decade ago. Because of all this, they will be able to intelligently absorb text on a screen (even within the form of a novel) alongside a myriad of other digital distractions, and it's an insult to them to say that they won't – not to mention that novels of the future will reflect and celebrate these changes, not provide an antidote to them. Yes, some people will continue to hug novels in bay windows on autumn days, basking in the warm glow of a fireplace with a cup of chamomile at their side. But many more will embrace the convenience and advanced usability that digital technology and electronic reading provides, and for them nothing will be lost in the equation.

And of course writers like Doctorow are not alone; the list of authors who resist their work being read on a screen is today surely longer than the list of those who would welcome it. J.K. Rowling is yet another author who believes that the only place for words is on the page. Because of this, she has resolutely insisted that each of her blockbuster Harry Potter books not appear in digital form. Of course, for someone who writes long books in longhand, not to mention spending all of her time in a fantasy world, her viewpoint is understandable if not predictable.

However, if the Harry Potter books were made available as eBooks the sales for them would be huge and it would put a stop to the widespread – and meticulously coordinated – piracy efforts that are always put into effect seconds after each new book in the series becomes available. What Rowling doesn't seem to realize is that people who want to read her book electronically are going to do so anyway, so why not let them do it legally? People would never bother to pirate something that already exists (which the success of iTunes has proven). But Rowling seems to be desperately clinging to some Victorian notion of a writer as a scribbler of handwritten tomes, noble ink-stained fingers making delicate row upon row of script on foolscap. Meanwhile, studies have shown that – despite the crazed interest in her books – the reading habits of kids are in serious decline; they're spending much more time with computers than they do with books. If we could get them to read the Harry Potter books electronically, it could begin to get them into the habit of merging the reading of text with the use of computers, and it would at least be a chance to reverse some very serious trends in terms of youth illiteracy. But instead, Rowling and others cling to their pads of paper in an iPod world.

Booth Tarkington's 1918 Pulitzer Prize-winning novel, *The Magnificent Ambersons*, takes place at a time of immense cultural change: the dawn of the twentieth-century's industrial revolution. Mechanization was replacing the handiwork of men, and many feared that machines would one day take over all aspects of society. At the center of the book is a battle of wills between the obnoxious and insufferable George Minafer, a young man who is part of the distinguished Amberson family that owns much of the small Midwestern town where the story takes place, and Eugene Morgan, an automobile manufacturer who was George's mother's first love. One night when Eugene is dining with the Minafers, George – never one to hold his tongue – blurts out, 'Automobiles are a nuisance… They'll never amount to anything but a nuisance.' Since Eugene has staked his fortune and reputation on the future of the automobile, there's of course tension in the room. (This scene is also included in Orson Welles's brilliant 1942 film version, with Joseph Cotten reciting most of the following speech.) Much to the amazement of those around the table, instead of rising to the defense of the automobile, Eugene agrees with George. 'I'm not sure he's wrong about automobiles,' he says. Eugene continues:

> With all their speed forward [automobiles] may be a step backward in civilization – that is, in spiritual civilization. It may be that they will not add to the beauty of the world, nor to the life of men's souls. I am not sure. But automobiles have come, and they bring a greater change in our life than most of us suspect. They are here, and almost all outward things are going to be different because of what they bring. They are going to alter war, they are going to alter peace. I think men's minds are going to be changed in subtle ways because of automobiles; just how, though, I could hardly guess. But you can't have the immense outward changes that they will cause without some inward ones, and it may be that George is right, and that the spiritual alteration will be bad for us.

The same could be said about what will happen in terms of the disappearance of books and the emergence of digital reading. Many other print-related industries, such as magazines and newspapers, have already experienced just how hard this transformation can be. Thousands of jobs have been lost, and the success and failure of entire companies have been decided by the question of whether they embrace digital change or else how much they try to resist it. Publishing has so far escaped close digital scrutiny, but this reprieve cannot last much longer. Publishing – one of the oldest forms of commoditized entertainment that exists – is going to have to realize that change is coming, and no one (not the booklovers or the book industry) can stop it any more than previous generations a hundred years ago could stop the ascendance of the automobile.

2

us and them

THERE CAN BE no doubt that we live in a digital world, and that the influence of computers and the Internet now play an important part in everyday life. Yet there's a very real and widening gulf between bibliophiles and those who preach the new gospel of electronic change. But while this may seem like a very new argument, filled with zeitgeist terms and modern day gadgets, at its heart is a conflict that is – at the very least – already a half-century old.

In a controversial 1959 speech entitled *The Two Cultures*, British writer C. P. Snow described how the sciences and the arts and humanities were at that moment being increasingly segregated into separate camps by a growing and profound split in thinking and values. Each group distrusted the other, with the artists looking at the scientists as if they were boorish philistines, while the scientists regarded the artists as clueless Luddites.

'The non-scientists have a rooted impression that the scientists are shallowly optimistic, unaware of man's condition,' wrote Snow. 'On the other hand, the scientists believe that the literary intellectuals are totally lacking in foresight, peculiarly unconcerned with their brother men, in a deep sense anti-intellectual, anxious to resist both art and thought to the existential moment.'

At the time, Snow – a scientist by training who turned to writing later in life – was commenting on the rift he saw enveloping London throughout the 1950s. Yet the atmosphere he described sounds very much like one which reappeared in force at the dawn of the twenty-first century when a growing digital society began to encroach upon almost all aspects of life. The only real change in Snow's scenario is that the scientific culture involved is no longer men in white lab coats trying to split the atom, but is instead computer scientists and inventors who are reimagining and rewiring our daily lives. Other than that, Snow's criticism and predictions seem remarkably fresh. For instance, here's a passage from Snow's original essay which appeared in the *New Statesman* in 1956; when examined in light of the debate about the future or the relevance of the book, Snow's words seem incredibly apt:

> The traditional culture, which is, of course, mainly literary, is behaving like a state whose power is rapidly declining – standing on its precarious dignity, spending far too much energy on Alexandrian intricacies, occasionally letting fly in fits of aggressive pique quite beyond its means, too much on the defensive to show any generous imagination to the forces which must inevitably reshape it.

The behavior to which Snow is referring can be seen in dozens of arguments that bibliophiles make in defense of the book, with John Updike going so far as to claim in a 2006 speech to booksellers that 'books are intrinsic to our human identity.' *The Beverly Hills Diet* is a book. Does that mean *The Beverly Hills Diet* is intrinsic to our humanity?

'The clash is between what you might call the technorati and the literati,' wrote Bob Thompson in *The Washington Post* shortly after Updike's speech. 'The technorati are thrilled at the way computers and the Internet are revolutionizing the world of books. The literati fear that, amid the revolutionary fervor, crucial institutions and core values will be guillotined.'

This fear is felt throughout the discussion over the fate of the book. Critics and writers cling to the status quo, while booklovers are reluctant to say goodbye to the book and refuse to believe that digital delivery and consumption of text could somehow be a good thing.

'Persons ignorant of the nature of change, antagonistic to the scientific revolution which will impose social changes such as none of us can foresee,' wrote Snow in *The Two Cultures*, 'often think and talk and hope as though all literary judgments for ever will be made from the same viewpoint as that of contemporary London or New York: as though we had reached a kind of social plateau which is the final resting-ground of literate man. That, of course, is absurd.'

Unfortunately, the level of snobbery that Snow describes is still evident in the debate about the future of the book. More than a decade ago Pulitzer Prize-winning author E. Annie Proulx stated in *The New York Times* that the Internet was good for 'bulletin boards on esoteric subjects, reference works, lists and news – timely, utilitarian information, efficiently pulled through the wires.' Sounds reasonable enough, until she adds that 'nobody is going to sit down and read a novel on a twitchy little screen. Ever.'

Proulx has been initially proven correct, at least in terms of wide-scale adoption, because people *aren't* reading electronic books in large numbers. But that's a very different thing than saying that people aren't reading *anything* online. Because if you factor in things like email, social networking websites, blogs, and wikis, people now read probably more than they ever did. But writers like Updike and Proulx continue to make the argument about the book, and not reading, ignoring the larger implications.

Behind these various arguments and statements, it's very easy to see the edge of what Snow was describing fifty years ago. While computers are involved in almost every aspect of everyday life – there's still a real distrust by intellectuals of technology. The gap Snow wrote about in *The Two Cultures* is only expanding, not shrinking. Books have now moved to the center of this debate, and the stakes couldn't be higher.

Ray Bradbury's *Fahrenheit 451*, written just a few years before *The Two Cultures*, tells the story of a world in which print is not only dead but is also extinct and illegal. The book's main character, Guy Montag, is a fireman in a world that has already seen two atomic wars and is on the verge of a third. In this post-apocalyptic reality, firemen start fires instead of putting them out. Specifically, Montag and his fellow firemen burn books, along with the houses in which they are found.

He describes his job early in the novel to a young girl: 'It's fine work. Monday burn Millay, Wednesday Whitman, Friday Faulkner, burn 'em to ashes, then burn the ashes. That's our slogan.' And instead of society being up in arms and trying to prevent this, firemen are held in some esteem. No one much minds the loss of books in their lives. Montag, however, begins to slowly question his work and that of his fellow firemen. He starts to wonder about the power books possess to be at once so feared by the state and so revered by those who risk their lives by owning them.

After a particularly harrowing night at the house of a woman who chooses to die by setting herself on fire rather than face the prospect of living life without her books, Montag talks to his wife, Mildred, trying to understand why people would be willing to take such risks and face such consequences. He decides, 'There must be something in books, things we can't imagine, to make a woman stay in a burning house; there must be something there. You don't stay for nothing.'

Shortly after this, disillusioned and looking for answers, Montag tracks down a strange old man named Faber he'd met the previous year. Faber turns out to be a professor who hasn't been able to teach for forty years (all teachers and educators are similarly unemployed, or have gone underground, since the texts they used to teach are now forbidden). Faber attempts to talk sense into Montag by convincing him that 'it's not books you need, it's some of the things that once were in books.' Montag doesn't seem to understand, and wants to risk his life for these inanimate objects that, until recently, he'd never held in his hands except to toss onto a raging fire. Faber tries again to tell him that it's what books contain – the ideas, the language, the stories – that are really what he's after:

> No, no, it's not books at all you're looking for! Take it where you can find it, in old phonograph records, old motion pictures, and in old friends; look for it in nature and look for it in yourself. Books were only one type of receptacle where we stored a lot of things we were afraid we might forget. There is nothing magical in them at all. The magic is only in what books say...

Bradbury's vision of a world without books was more than just something on which to hang his plot; it was a reaction to what he saw happening around him in 1950s America. Hollywood's Technicolor fantasies were incredibly popular, and formulaic television was on the rise; reading had begun its slow decline. And while many features of his novel are indeed prescient (the moronic television shows that Montag's wife is addicted to sound an awful lot like the reality shows that are on the air today), the book is – like most good science fiction – a cautionary tale more than a prediction of things to come.

We can glean many things about the nature of books from Bradbury's novel, and our reaction to them, not to mention what their true worth is and why we should care about them (and, more importantly, why we should care about what's *in* books more than the books themselves). But what's most important to note about the scenario of *Fahrenheit 451* is that – in the novel's alternate reality – books were not initially banned by the government. They didn't need to be banned. People just stopped reading, turning instead to other forms of entertainment. So when the state finally got around to making the ownership of books illegal, no one much missed them or even cared.

Bradbury was not worried about the loss of books themselves. His main concern was with a society that killed the need for reading, replacing the act with the narcotic of passive entertainment. The disappearance of books was merely what happened when people stopped caring about them. Unfortunately, according to a landmark 2004 study conducted by the National Endowment for the Arts entitled *Reading at Risk: A Survey of Literary Reading in America*, this is exactly what's happening.

Reading at Risk – the result of a national survey spanning 20 years, and ultimately providing 'so much data in such detail that it constitutes a comprehensive factual basis for any informed discussion of current American reading habits,' – was heralded as a frightening and important wake-up call when it was first published. What the survey found was, in its own words, 'a bleak assessment of the decline of reading's role in the nation's culture.'

In his preface to the study, NEA chairman Dana Gioia wrote that:

> ... for the first time in history, less than half of the adult population now reads literature, and these trends reflect a larger decline in other sorts of reading. Anyone who loves literature or values the cultural, intellectual, and political importance of active and engaged literacy in American society will respond to this report with grave concern.

Indeed, Gioia summed up the report in two sentences:

> Literary reading in America is not only declining rapidly among all groups, but the rate of decline has accelerated, especially among the young. The concerned citizen in search of good news about American literary culture will study the pages of this report in vain.

It's also important to note that, in the NEA report, the word 'literature' is used in broad terms; the study is not mourning the fact that kids are reading Tom Wolfe instead of Thomas Wolfe. Instead, the report found that kids aren't reading much of anything, choosing instead to play videogames and spend their time cruising the Internet instead of picking up a book, newspaper or even a magazine.

The report is also not from the point of view of tenured professors preferring that the youth of America – or anyone, really – read Shakespeare or Racine; it is instead an in-depth report showing that almost every type of reading is in freefall among youths (not to mention that book reading is declining significantly even for adults).

While the report tries to stick to the facts, choosing to feature chart after chart of data instead of paragraphs of theory or conjecture, it does indeed offer tantalizing bits that show what books are up against: 'Literature now competes with an enormous array of electronic media. While no single activity is responsible for the decline of reading, the cumulative presence and availability of these alternatives have increasingly drawn Americans away from reading.'

The study also places the survey in historical context, noting that the decline in reading has occurred during the rise of inventions and phenomena such as videogames and the Internet. But again, it tries its best not to draw parallels. Instead, '*Reading at Risk* merely documents and quantifies a huge cultural transformation that most Americans have already noted – our society's massive shift toward electronic media for entertainment and media.'

What of course makes this study most important is the warning it should provide to the book industry (publishers, editors and even writers) who need to engage this new generation which is beginning to turn its back on books. The time is especially ripe since, according to *Reading at Risk*, in the first decade of this new century we are now 'at a critical time, when electronic media are becoming the dominant influence in young people's worlds.' According to this survey and its admittedly 'bleak assessment,' it would seem that American society is in the first stage of what Bradbury pointed to in *Fahrenheit 451*: books going away because people simply stop caring about them.

In the end, it really won't be software or technology that kills books or print. It won't be the Internet or some organized challenge or even attack upon the kingdom of reading. Print will disappear merely because of – like in Bradbury's brilliant novel – lack of interest. Because no one wanted – or most people just forgot about – books. To paraphrase Nietzsche's famous edict, 'God is dead, and we killed him,' I would say that print is dead, and the Internet killed it.

In the same way that modern society, a hundred years ago, erased the need for God in the everyday lives of its citizens, thereby 'killing God,' the invention of the Internet is killing print little by little by removing it as a necessity for most people. Information and news are rapidly spread by websites, and most people communicate via emails. Sports scores, apartment listings and classified ads used to be found in newspapers, but most people now get this information online. The need for print is slowly disappearing, at least on a grand scale, and thus it is slowly being erased.

In terms of Nietzsche, his theory was just a metaphor, and some would claim not a very effective one. After all, people today still believe in God. However, God is no longer the force he once was, controlling people's lives as well as the actions of governments or continents (save, unfortunately, for the demented rise in religious fundamentalism and fanaticism). Books could be headed for a similar fate; they'll be on our shelves, but no longer in our hearts and minds.

The world of literature used to be a place where new ideas were formed and exchanged, so when did it start wearing blinders? And why is it so reluctant to change?

The main reason is because change is always a risk, and book publishers – who have managed to hold their own against all of the other potential distractions to their collective bottom line over the past fifty years – have grown a bit cocky in their belief of their product. This goes beyond an 'if it ain't broke, don't fix it' mentality. Instead, many of those in publishing see themselves as guardians of a grand and noble tradition, so much so that they sometimes suffer delusions of grandeur.

'Unlike other industries, our product is a book, arguably the highest form of human endeavor,' wrote Pat Walsh in his 2005 book *78 Reasons Why Your Book May Never Be Published and 14 Reasons Why it Just Might*. 'The book is the conduit by which most of the greatest minds ever known have chosen to, or been compelled to, communicate with the world.'

The scariest thing about the above is that I think he's serious. Many others in publishing feel the same way. And while this is at worst pretentious or self-important (again, remember: *The Beverly Hills Diet*), it's also dangerous. Because when a person, or even an entire industry, is convinced that they're on a divine mission, then they're going to be even less, rather than *more*, inclined to change their ways and try something new. And the more that publishers ignore the drive (much less temptation) to change, the more they will marginalize themselves, pulling away from the rest of the culture, and instilling an *us or them* worldview that may take years to shake.

Also, what's ironic about Walsh's idea – which unfortunately is shared by thousands of others in the industry – is that books have indeed been used as the primary tool for communication in the past. But not many people in the industry are willing to admit or acknowledge that the Internet is now the prime vehicle for the dissemination of information. For everything from online news to Wikipedia (with Google tying it all together), the Web is where people go when they're looking to gain access to content. This change has already happened, and publishing now needs to react instead of preach.

'For if, like immature children, we steadfastly maintain our allegiance to the obsolete institutions of the past, then we will certainly go down with the ship,' wrote Douglas Rushkoff in his 1999 book *Playing the Future: What We Can Learn from Digital Kids*. 'On the other hand, if we can come to understand this tumultuous period of change as a natural phase in the development of new kinds of intelligence and cross-cultural intimacy, then our imaginative and creative abilities are the only limits on our capacity for adaptation.'

And it's not just booklovers who need to understand that times are changing. Consumers are going to have to also change. While the failure of eBooks (of which more later) shows that there is indeed resistance to the idea of wide-scale digital reading, it also shows that consumers are not interested in replacing books when there's not a viable alternative. Because even though there's an abundance of ways to read material over the Web – in terms of blogs and online news – there is not yet a good way to read narrative fiction and lengthy non-fiction books. The mass consumption of books has not yet been replaced by the Internet, but it's only a matter of time.

'Users will change their habits,' wrote Pip Coburn in his 2006 book *The Change Function: Why Some Technologies Take Off and Others Crash and Burn*, 'when the pain of their current situation is greater than their perceived pain of adopting a possible solution.'

The problem with wide-scale adoption is that, right now, there's no real pain associated with reading books. They're easy to find, relatively inexpensive, and there's a great selection of them available. Additionally, in terms of digital reading, there's still confusion on behalf of the consumer, as well as a number of problems with the eBook business model – in terms of pricing, selection, formats and digital rights management – which continue to make them unattractive. Integration with already existing devices, such as Tablet PCs or Apple's iPhone, will make it much easier for consumers to make the decision between print books and electronic books. But even at that point, digital reading will take some getting used to.

This is not to say that change will be easy, either for the consumer or for the industry, but as the success of the iPod has recently shown, people (and industries) are very quick to adapt to new ways of delivering and consuming entertainment.

'No one thought the iPod would change the music business, not only the means of distribution but even the strategies people would use to buy songs,' wrote Steven Levy in his 2006 book *The Perfect Thing: How the iPod Shuffles Commerce, Culture, and Coolness*. 'No one envisioned subway cars and airplane cabins and street corners and school lounges and fitness centers where vast swathes of humanity would separate themselves from the bonds of reality via the White Earbud Express.'

Now, merely five years after the introduction of the iPod, the device is not only nearly ubiquitous in the minds of consumers but it has also led the way in changing an industry which had not faced a serious threat to its existence in decades.

'The absolute transformation of everything that we ever thought about music will take place within 10 years, and nothing is going to be able to stop it,' said David Bowie in an interview with *The New York Times* in 2002, acknowledging the changes that the Internet had brought:

> I see absolutely no point in pretending that it's not going to happen. I'm fully confident that copyright, for instance, will no longer exist in 10 years, and authorship and intellectual property is in for such a bashing.... It's terribly exciting. But on the other hand it doesn't matter if you think it's exciting or not; it's what's going to happen.

Besides, we've seen this level of change before. Five hundred years ago, when books were first introduced, they were greeted with the same level of skepticism that digital reading is facing today. Gutenberg's bibles, as much as we revere them now, were not welcomed with open arms or eager hands.

'Medieval clerics greeted printed books as imposters of illuminated manuscripts – aesthetically inferior, textually unreliable and likely to breed a dangerous diversity of opinion,' wrote Jacob Weisberg in *The New York Times* in 2000. 'The echo of such views is heard today in an equally misguided elite's hostility toward digital publishing.'

Even though the idea of digital reading has been around for decades the hostility persists, the very idea of electronic books making people uncomfortable. 'We are today as far into the electric age as the Elizabethans had advanced into the typographical and mechanical age,' wrote Marshall McLuhan in his 1962 book *The Gutenberg Galaxy: The Making of Typo-graphic Man*. 'And we are experiencing the same confusions and indecisions which they had felt when living simultaneously in two contrasted forms of society and experience.'

Whereas C. P. Snow wrote about the two cultures, McLuhan was getting down to the nitty gritty of the argument, describing the actual clash of experiences. True, the new is very different from the old, and to make that change will be difficult. But *difficult* doesn't mean it won't happen.

What the critics of digital reading fail to realize is that it already *has* happened; people have already made substantial changes in their daily lives when it comes to digital reading. What do these critics think happens with text messages, searching for an apartment on craigslist, not to mention the hundreds and thousands of blogs that have become required daily reading by millions? What about the workplace, where employees spend their days glued to their computer screens looking at memos, documents, reports and email?

What's happening in all of these examples is the same: reading. And while that may not be snuggling up to Stendhal on a park bench on a crisp Fall day, it is still reading. It's still words being taken in on a computer screen, and for millions of people it is a daily occurrence, one that now seems as natural to them as anything else in their lives. To think that millions won't be willing or able to make the transition to an overall digital reading experience is naïve. In large measure, people already do the majority of their reading digitally.

In the end, we may be in love with books, but it's words that have truly won our hearts. It's words that whisper into our ear and transform us, that make us believe in other worlds or new emotions we didn't know existed; it's words that keep us company in those planes, on subway trains, or our comfy couches. It is words, not books, paper, papyrus or vellum pages that transform our lives. Joseph Stalin, knowing the sheer power writers possessed in shaping ideas and changing minds, called them, 'The engineers of human souls.' Stalin wasn't afraid of books themselves – inert pieces of pulp and cardboard – he was afraid of the ideas that books contained. Adolf Hitler's bonfires were much the same; he wasn't trying to eradicate books; rather the ideas that they contained. Hitler was, after all, himself an author who had used a book to spread his ideas.

It is not the ardent booklovers that the NEA survey is discussing or addressing; indeed, the ones who have stopped reading probably have also not bothered to read the report. So the people who repeatedly strike their chests in defense of books are seriously missing the larger battle, which is the rising tide of apathy and lack of interest towards reading. This is where the real war is; this is what's important. Yet, instead of addressing this very real concern, critics of digital reading or electronic books merely extol the aesthetics (and not necessarily the virtues) of printed books.

In a *New York Times* story from 2006 entitled 'Digital Publishing is Scrambling the Industry's Rules,' author Anne Fadiman is quoted as saying, 'For reading, you have to read a book in its entirety and I think there's no substitute for the look and feel and smell of a real book – the magic of the paper and thread and glue.'

There are many similar odes to books and their physical nature; how much their owners like to hold them and stroke them (as if they were pets instead of messengers conveying ideas). It really is amazing that people can't seem to intellectually separate their love for reading from their love of books.

But people have felt this way about numerous commodities that have since become endangered species. We heard about this in the 1980s when CDs began to rub out vinyl, about how losing the snap and crackle of the needle slipping into the groove meant losing the soul of the music. How the voice of Dylan or Lennon should never be separated from its true sinewy waves, to be instead reduced to jagged lines represented by zeroes and ones. And to debate this in the context of listening to music is one thing, but would a sane person say that they'd rather people listened to vinyl than not hear *Blonde on Blonde* or *Rubber Soul* at all?

So to insist on saving just books – treating them as if they were mere props in the movie of our everyday lives, destined to sit on a shelf and just look nice – as many people think we should do, is foolish and ignores the bigger problem.

We need to realize that we live in a time of almost unimaginable change, and to think we can have such transformation in other areas of our lives but have books and publishing stay the same, is naïve bordering on irresponsible. And of course, for books to change, the business models on which the industry of publishing has been built for the last century will also have to change.

While those in publishing hem and haw and wearily engage in this debate at various levels, an entire generation has already decided that print is dead. Indeed, for them – raised on the Internet – it might not ever have been alive.

'But books not only define lives, civilizations, and collective identities,' wrote Nicholas A. Basbanes in his 2005 book *Every Book Its Reader: The Power of the Printed Word to Stir the Soul*. 'They also have the power to shape events and nudge the course of history, and they do it in countless ways.'

In those instances where books changed societies it was because those societies were best reached by books; you can reach more people through a book that can be printed over and over than you can standing on the street trying to get people to listen to your point of view. And yet, in this analogy, the street corner is no more important or better than the book; they both exist as a means to convey the point of view or the idea. Today, the Internet would be the best tool to quickly communicate with a large number of people.

What should be remembered at all times is that the words compiled into books have a much larger purpose than to collect dust on shelves. It is ideas that matter and should be unleashed, not constrained by print. The Declaration of Independence changed the history of America not because it comprised written words on a page, but because it was the physical embodiment of philosophy and ideas. We celebrate the moment in time the paper represents; we don't worship the paper simply because it's *paper*.

Not even the sleekest futurist who believes that one day all our food will be eaten in pill form and we will soon commute via buzzing hovercraft thinks that books should or ever will be completely banned or eradicated. Instead, what the proponents of digital reading are advocating is that literary content and text adapt to our increasingly electronic future and lifestyles. And, if it doesn't, then people won't only turn away from books but they'll also turn away from the stories and ideas found *inside* books. According to *Reading at Risk*:

> as more Americans lose this capability [to read], our nation becomes less informed, active, and independent-minded. These are not qualities that a free, innovative, or productive society can afford to lose.

It is these qualities we should be afraid of losing, not books themselves.

Will books last for hundreds of years? Yes, of course; no one is calling for bonfires. Will books *matter* in a hundred years? To

turn again to *Fahrenheit 451*, and the education of the disillusioned Montag on the importance of books, Faber tells him,
'Give a man a few lines of verse and he thinks he's the Lord of
all Creation. You think you can walk on water with your books.
Well, the world can get by just fine without them.'

3

newspapers are no longer news

IN THE CAMPY Ed Wood film *Plan 9 From Outer Space*, a couple are sitting on their patio staring up at the night sky. All of a sudden they're awash in bright light and surrounded by strange noises. They're both momentarily puzzled, but soon get over it, the husband laconically telling the wife, 'We'll find out what it was tomorrow in the paper.'

For a long time – for most of the past couple of centuries, in fact – people were content to do just that: wait until their morning newspaper to find out what had gone on in the world the day before. But the rapid rise, and now the global inter-connectedness, of the Internet has shattered that idea. People now want news in almost real time, getting updates as events are actually happening. Any news in a newspaper will be out of date by the time it's printed; it will in fact no longer be *news*. To quote TV comedian and political satirist Stephen Colbert, '*USA Today* should really be called *USA Yesterday*.'

In journalism, content really is king; what matters is the information. Few would argue for the aesthetic merits of an average newspaper. The cheap pulp is simply a vehicle for transporting the news and events of the day, a vessel into which is poured 'all the news that's fit to print.'

Cable television, with the invention of channels like CNN, discovered long ago that there was a lust for news twenty-four hours a day. And those channels, and many others like them, have now existed and prospered for decades by intravenously feeding news junkies constant information through their cable hookups.

Most big newspapers have elaborate websites where stories are posted as soon as they're written, as events occur. The home page of *The New York Times* is almost like a blog because it gets updated every few minutes, with the most recent stories located at the top of the page. It's easy to spot the recently updated articles because they have red tags underneath them, showing when they were last updated. Sometimes you can log on to the *Times* website and read a story that was posted only a minute or two before.

'As one industry after another looks at itself in the mirror and asks about its future in a digital world,' wrote Nicholas Negroponte in his 1996 book *Being Digital*, 'that future is driven almost 100 percent by the ability of that company's product or services to be rendered in digital form.'

Despite all of this, much of the newspaper industry has decided to resist the opportunity offered by digital delivery and consumption. Instead, they have mostly ignored or resisted the changes happening all around them in other media. And yet, on the production side of things, computers and technology have long played a part in the composing of news stories and the creation of the newspaper itself. However, these production changes were nothing compared to what the future would bring.

'Panel after panel was held at journalism conventions about whether newspapers would be replaced by the downloading of the day's news onto a computer screen,' wrote Anna Quindlen in her 1998 book *How Books Changed My Life*. 'It seemed only sensible to those whose correspondence had become characters sent by modem from one computer to another instead of a file of business letters, inevitable that the collection of folded newsprint that landed on the doormat with a *thwap* before daybreak each morning could simply be replaced by a virtual newspaper in a computer in the kitchen, coffee cup beside the keyboard.'

This is indeed the reality for millions of people every day: they still read the news, but they no longer have to go to a newspaper to get it. The Internet and digital delivery has exploded the idea of how news is gathered and distributed. Using the syndication of RSS feeds ('RSS' standing for Rich Site Summary, or Really Simple Syndication, depending on who you talk to), the news comes directly to consumers.

RSS allows readers to sign up to receive specific portions of individual publications, thus making it even easier for them to get the information they want. And still many in the newspaper industry are resisting any kind of real change to their product. Many deny that anything is wrong at all, and are spending lots of money to convince others that the sky is not falling.

In February of 2007 the Newspaper Association of America unveiled a $75 million dollar ad campaign which was 'designed to "surprise advertisers with the truth" about consumer engagement with newspaper advertising as well as the strength and vitality of the audience delivered by newspaper media.' The way they were going to do this was with print ads. The ads had a list of the variety of ways that newspapers were slicing and dicing their content for electronic consumption, without ever really mentioning that lots of consumers are bypassing traditional news outlets in general (not to mention newspapers themselves) and are getting their news from blogs and/or other online sources. In fact, the NAA seemed to think that news was created so that newspapers would have something to fill its pages, rather than the other way around.

To make matters worse, the graphic that they came up with to represent how news can be mashed into all kinds of electronic formats showed a rather creepy person completely wired with all sorts of gadgets stuck and attached to his head and back. The subtext of a person carrying so much digital baggage (and junk) seemed to subtly be – or really, not so subtly – 'Hey, buddy, forget all those wires and cords, and just pick up a good ol' fashioned newspaper from the newsstand.' I almost expected to see steam coming off the guy's coal-powered jetpack.

While all of this is going on, there has been news story after news story about the decline of print, with new studies and statistics showing that consumers are increasingly heading online for their news and information.

'U.S. newspaper publishers have been fighting to hold on to advertisers as many of them lose readers to other media, including the Internet,' stated a Reuters article from October 2006. 'Print readership fell, according to a comparison of figures from the two periods conducted by Reuters. *The New York Times* readership dropped 5.8% to more than 4.7 million people, while the largest U.S. paper, Gannett's *USA Today*, fell 3% to about 6.9 million.'

All of this is taking place while newspapers' profits and readership are in decline. Traditional readers of newspapers are going elsewhere for their news (mainly to online sources), while younger generations never got in the habit of reading newspapers in the first place. After all, why would they? What can *The New York Times*, sitting inert on a newsstand where it's been for hours, tell them about what's happening out in the world *right now*?

'Now the problem is to get people under 50 or so to pick up a newspaper,' wrote Michael Kinsley in a 2006 column entitled 'Do Newspapers Have a Future?' which appeared in *Time* magazine:

> Damp or encased in plastic bags, or both, and planted in the bushes outside where it's cold, full of news that is cold too because it has been sitting around for hours, the home-delivered newspaper is an archaic object. Who needs it? You can sit down at your laptop and enjoy that same newspaper or any other newspaper in the world.

More and more people are realizing this, and are turning their backs on newspapers. The newspaper readership numbers are beginning to reflect this. And yet, with all of this change taking place, many newspapers still refuse to budge. Why?

'The origins of the newspaper industry's rapid transformation can be traced to 1945,' wrote Elizabeth M. Neiva in the journal *Business and Economic History* in 1995. 'Before that time, the industry had enjoyed nearly seventy years of relative stability. There were no significantly technological innovations, few new competitive threats, and only minor cost increases. In the words of one publisher, "The whole industry simply coasted through the first half of the twentieth century."' After all of that coasting, the industry doesn't seem to want to come to terms with or even acknowledge its recent skid.

Book publishing is in a similar situation. While there have indeed been challenges and competition over the years – from radio and motion pictures and then television – none of these managed to make much of a dent. Most recently, the failure of eBooks since their introduction in the late 1990s has only given publishing more reason to stay on its current path and not change the way it does business. But the resiliency of major trade publishing is slowly hardening into hubris – the same kind of hubris that the newspaper industry has been cultivating for the past fifty years – and publishing is beginning to think that any challenge to the way it does business is an attack of philistinism rather than an idea whose time might have come. While some of this is tied to the general fear of technology (as C.P. Snow wrote in *The Two Cultures*, 'Intellectuals, in particular literary intellectuals, are natural Luddites'), some of it is just the usual arrogance that comes from ruling the roost for too long.

'By the end of the twentieth century, when virtually every publication in America was screaming like a carnival barker to hawk the wonders of cyberspace and the promise of friction-free commerce,' wrote Steven Levy in his 2006 book *The Per-*

fect Thing, 'your fear of change would have had to be very substantial indeed to limit your vision to the Internet's threats and not actively pursue its benefits.'

That fear of change, among publishers of all kinds of print products – magazines, newspapers, and books – is now so palpable that it's almost paralyzing. And yet the problem is the same with all of these businesses: they think that they sell products instead of entertainment, information or escape. This is why they're scared of what a digital future will bring.

After all, if electronic reading means the instantaneous downloading of text – probably through an instant wireless connection from anywhere in the world – then publishers are put out of business, right? If the physical object goes away, then surely the company who makes the physical object will also disappear. But this is of course wrong. Publishers in fact aren't in the magazine, newspaper or book business (in the sense of these things as physical objects); they're in the *idea* and *story* business. The only ones truly in the book, newspaper or magazine business are the paper mills and printing plants producing these products.

While newspapers in general have been under attack, in early 2007 a new kind of attack began to take place: a nearly unprecedented assault on book reviews. It began when a number of newspapers, among them *The Chicago Tribune, Newsday, The Minneapolis Star Tribune, The Memphis Commercial Appeal, The Cleveland Plain Dealer, The Dallas Morning News, The Sun Sentinel* and many others began to either shrink their book review sections or else get rid of them altogether.

But when *The Atlanta Journal-Constitution* made the decision to fire Teresa Weaver, its book review editor, the critics started to unite and the blogosphere exploded. The National Book Critics Circle Board of Directors, also known as the NBCC, launched a petition to save Weaver's job, and then proceeded to mount an all out 'campaign to save book reviews.' On their blog, in April of 2007, John Freeman – president of the NBCC – posted their manifesto:

> We're tired of watching individual voices from local communities passed over for wire copy. We're tired of book editors with decades of experience shown the exit so that the book section can be passed like a hot potato with no dressing. We're tired of shrinking reviews. We're tired of hearing newspapers fret and worry over the future of print while they dismantle the section of the paper which deals most closely with the two things which have kept them alive since the dawn of printing presses: the public's hunger for knowledge and the written word.

This attitude struck many as disingenuous. After all, where was all this fire when independent bookstores were going out of business? One by one, small bookstores across the country were decimated by the big chains, and never was a peep heard from critics. Why? Perhaps because Barnes & Noble still sold the books the critics reviewed. But once it was the reviewers themselves that found themselves in the crosshairs, it started to get personal. Also, what went unmentioned by the NBCC was that the *Atlanta Journal-Constitution* was laying off a number of writers, including critics covering classical music and visual arts. Why no mention of them?

Following the launch of the NBCC campaign and its website, writer Art Winslow had an essay on the Huffington Post website entitled 'The New Book Burning.' Wrote Winslow:

> In the new book burning we don't burn books, we burn discussion of them instead. I am referring to the ongoing collapse of book review sections at American newspapers, which has accelerated in recent months, an intellectual brownout in progress that is beginning to look like a rolling blackout instead.

Winslow was more than slightly hysterical when he tried to portray the disappearance of book review sections as being 'the new book burning.' That's not only a ridiculous suggestion, but a dangerous one. Burning books is about the totalitarian eradication of what the ideas in books represent, whereas book review sections being slimmed down or phased out is about simple economics and the fact that things are rapidly changing and book reviews are no longer needed. But Winslow prefered to take a darker view, rhetorically asking, 'How did we arrive at what seems to be a cultural sinkhole?' Instead of answering that, I'd like to ask Winslow where he's been for the last ten years.

What I find most interesting about Winslow's essay is that he's a 'former literary editor and executive editor of *The Nation* magazine and a regular contributor to *The Los Angeles Times*, *Chicago Tribune*, *Bookforum* and other publications.' So it seems that Winslow, and many critics and writers like him, are really just clamoring to keep their jobs. In the end, they don't want things to change because they don't want to give up the power they currently have.

Winslow and other book reviewers used to act as the arbiters of literary taste: when they would write a good review of a book, their review had the power to propel that book into the national spotlight (and vice versa; a bad review could ruin a book, and sometimes an entire career). So while the importance of movie critics has lessened over the years (gore-fests like *Hostel* and *Saw*, which are routinely savaged by reviewers, go on to make millions at the box office despite what any critic says), in the book world, reviewers have – until fairly recently – retained their clout. But with the Internet, blogs, the rise of 'citizen journalism' and user-generated content, book reviewers are seeing their little corner of the world erode and fall into the sea, and they don't like it.

Even though Winslow and others fortified their arguments with the righteousness of fighting for culture, what they really can't stand is that things are changing and they're being left behind. Yes, book reviewing is an art, but that art is going away – the same way in which the skills that it took to produce a rotogravure or daguerreotype were also arts that disappeared. Things changed, the culture shifted; new machines were invented and new ideas were minted, and those skills went away. So while Winslow and others can lament the loss of book review sections in newspapers around the country, social networking sites like Library Thing, Shelfari and Good Reads are proving that literary discussion, sharing and discovery are still taking place.

When Winslow himself wrote that the loss of book review sections will '[choke] off such discussion of books,' he couldn't have been more wrong. There is now, because of the Web, probably more discussion of books than ever before. But what really infuriates Winslow and many of the other critics is that all of this discussion is happening without them. It's not that books are being burned; instead, what's happening is that the self-importance of book reviewers is going up in smoke.

Shortly after Winslow's piece appeared, bestselling mystery author Michael Connelly had an opinion piece in *The Los*

Angeles Times. Entitled 'The folly of downsizing book reviews,' it was yet another essay dealing with the closure and reduction of book review sections. While not as apocalyptic as the Winslow piece, Connelly used similarly dire language, stating that 'newspapers that cut back on book coverage may be cutting their own throats.' For Connelly, this was personal since he feels that it was positive reviews of his first book that saved (and gave him) his career. He then asked what would happen to a similar book in today's culture where book review sections are rapidly disappearing. However, Connelly happens to answer his own question with his opening sentence: 'Fifteen years ago, my first book was published in near obscurity.'

Within the past fifteen years the Internet was made available to the public, which itself has since given birth to dozens of new ideas and ways to communicate. The fact that, in the last fifteen years, we have been witness to the rise of blogs, user-generated content, iPods and MP3s shows that the world has changed substantially since the publication of Connelly's debut.

While book reviews saved his first novel a decade-and-a-half ago, the power to promote and form opinion has since shifted, moving away from print-based book reviews towards something much more egalitarian and open-ended. Because of the Internet, dozens of new ways to champion books now exist. Today a positive mention on the popular blog Boing Boing probably has the same power (if not more) to shape influence and spread the word about a book than a book review did back in the days of Connelly's first novel. And in addition to Boing Boing there are dozens of literary blogs, not to mention the various social networking websites devoted to books, all of which – cumulatively – have a much broader reach than book review sections ever did.

Former publishing executive and book review editor Pat Holt was next to jump into the debate, writing a great essay as part of her 'Holt Uncensored' email newsletter, entitled 'Book Critics: Are we driving readers away?' In it, Holt looked at the phenomenon not only of shrinking book review sections in American papers but, almost more significantly, the large number of mainly self-serving essays and think-pieces (not to mention the petition) that suddenly appeared in order to rally support for book reviewers and book review sections. Refusing to blindly jump on the 'We have to save book reviews!' bandwagon, Holt took an intelligent and thoughtful look at the situation, writing that 'maybe it's time for those of us who have worked as critics for a living to evaluate what's happened to our profession – and why we may be driving readers away. In the last 25 years, just about everything about the print experience has changed – except the way critics review books.'

Instead of the usual toothless arguments that the proponents of books and book reviews usually trot out in the 'print is dead' debate, Holt argued that the status quo isn't worth saving. Her whole point was that the world has changed; we have become more and more adept at finding information and content online, and now the literary world has to also change. 'Our audience zips around the Internet with tremendous agility and speed, and what do we give them?' asked Holt. 'Stodgy, dull, laborious and indulgent reviews.'

Holt also realized that the reduction of book reviews is only the tip of the iceberg, and that the loss of interest in book reviews is an early signal that a loss of interest in books themselves could be around the corner. But even that hasn't been enough to create change. 'Not only have we gotten stuffy, dreary and plodding, but our panic is showing – we know traditional print media is in trouble and try too hard to get readers back,' wrote Holt. 'We've substituted opinion for criticism. We've pronounced books good or bad rather than shown readers why.'

Even the book reviews that remain are no longer doing what book reviews used to do so well: connecting readers with books. Instead, readers have gone elsewhere, and are now connecting with reading material online. Without the aid of major newspapers or literary critics, consumers are finding new reading material from either recommendation websites or software on commerce sites, social networking sites, or even just by keeping in touch with people online who share with them ideas for books. So while a number of insulated critics are interested solely in protecting their turf (and their jobs), Holt's essay was asking for the industry as a whole to take a long hard look at itself.

The New York Times inevitably weighed in on the situation, with a story by Motoko Rich entitled 'Are Book Reviewers Out of Print?' Rich summed up the scene in a sane way, and then nicely summarized both the dilemma and the opportunity:

> To some authors and critics, these moves amount to yet one more nail in the coffin of literary culture. But some publishers and literary bloggers – not surprisingly – see it as an inevitable transition toward a new, more democratic literary landscape where anyone can comment on books.

Rich talked with a good selection of important bloggers, as well as publishing executives (who, somewhat surprisingly, thought that literary blogs were a good thing).

However, a writer who came off very poorly in the piece was the novelist Richard Ford, who denigrates book blogs without ever having read one. He even goes so far as to state that the *Atlanta Journal-Constitution* should print reviews 'as a public service.' This is the height of the literary establishment's vanity. Garbage collection and paved roads are a public service; what Michiko Kakutani thinks of D. B. C. Pierre is not. Even within the literary world, this isn't the loss that Ford and others make it out to be. To think that the world of literature will be worse off for not having a full-page photograph of Michael Chabon on the cover of *The Los Angeles Times Book Review* is lunacy; what is needed is for Michael Chabon to continue writing intelligent and entertaining novels. Meanwhile, the NBCC continued to relentlessly stage their campaign to save book reviews as if it were a push for civil rights or an anti-war rally. They even began staging 'read-ins' and protests, while their petition attracted the signature of people like Norman Mailer; at any moment I expect them to march on Washington and try to levitate the Library of Congress.

Unfortunately, all of these efforts are going to backfire, and will only show how out of touch the literary establishment is in terms of knowing who readers are, let alone knowing what they want. Instead of reaching out to readers, the reaction of the literary establishment has had the condescending whiff of 'let them eat cake' combined with 'father knows best,' while the critics themselves have circled their wagons in order to protect the status quo. But readers have been too busy, discovering books in dozens of ways besides a book review, to notice.

What's most ridiculous about all of this is that many reviews continue to appear; it's just that they're online instead of in the paper. But even that's not good enough for the critics. Why? Because, as NBCC president John Freeman states, 'you can't bring an online book page into the bath.'

This seems more silly than Margaret Atwood's claim about not being able to take books into the tub simply because most book reviews aren't immersive experiences. They're created expressly for the purpose of consumption in one sitting. Most reviews are tailor-made for digital delivery, since short pieces are easily consumed on handheld screens or laptops.

But Freeman seems to think that the fact that most book reviews appear online means that they somehow suffer from a lack of portability,' when it's actually exactly the other way around. Digital content can be accessed in a myriad of ways, on dozens of devices and gadgets anywhere in the world (not to mention that it can be available forever in archives). Paper is a perishable object bound to a single location that can be easily misplaced, ripped or stained, whereas content on a website is always there, forever unsullied and pristine, waiting for someone – anyone, anywhere – to touch a few keys and access its knowledge.

totally wired

4

generation download

IT BEGAN, AS lots of trouble does, in a dorm room. It was 1999, and Boston college student Shawn Fanning (nicknamed 'napster') wanted to find a way to trade and share music with his friends. Frustrated by the computer methods that were then available, which either took too long for songs to download or didn't allow him to get all the music he wanted, Fanning took matters into his own hands. He wrote a computer program that rather elegantly – and, as was finally decided, illegally – turned every user's machine into a server that other users could access. Instead of storing songs on one central server or machine, every person who became part of the network added their own libraries of music to the ever-growing catalog, until finally Napster (as Fanning eventually called his program) had an index of millions and millions of songs.

Fanning distributed the program for free, and word soon spread from his own Northwestern University to colleges all over the country. In no time college kids everywhere were downloading hundreds of thousands of songs a day from the network of computers that Napster made it incredibly easy to access. For college students – or anyone else – who was searching for music, and were also willing to share their own record collections, this was a fantastic development. But the people who downloaded almost every song, from the millions that were downloaded, avoided paying royalties to the copyright holders.

'Musicians were divided from the beginning,' wrote Joseph Menn in *All the Rave*, his 2003 history of Napster. 'Unknown acts saw the MP3 phenomenon as a way to spread their music. Brand-name acts, which had more to lose through piracy, were naturally more conservative. But even some of them wanted to release the occasional track digitally.'

While the artistic community debated the merits of Napster, the record labels that had spent millions signing groups and paying for their recording and touring costs were not quite so sanguine. Almost instantly Napster faced legal challenges, and the labels – teaming up with the Recording Industry Association of America – embarked on a disastrous campaign of suing individual Napster users (which sometimes included kids as young as twelve, in addition to senior citizens). Numerous cases of woefully bad PR followed, including the infamous incident in 2000 involving Metallica drummer Lars Ulrich, who personally delivered to the Napster offices the names of over 300,000 users who had been accused of illegally trading Metallica songs. Because of this, Ulrich quickly became the symbol of money-grubbing corporate greed, and the most hated man in rock and roll.

By 2001, it was all over. Napster was forced to shut down its network. German media conglomerate Bertelsmann acquired Napster in 2002, and since then the company has – mostly unsuccessfully – tried to rebrand itself as a legal way for people to buy and download music. But while Napster itself didn't last, it showed both kids (and corporations) the potential power of file-sharing.

What's also interesting is that Fanning was a Digital Native, a young kid for whom the building blocks of BASIC were his Legos. Whereas other generations spent time mischievously crank calling neighbors or local businesses, Fanning, along with his friends, were writing computer code that would eventually change the world. And all of this came about because they wanted to share and discover music.

This had been going on before Napster, but in much more (literally) pedestrian ways. For instance, I distinctly remember, as a kid, sharing music by meeting up with friends carrying Led Zeppelin and Rush LPs under my arm. In exchange they gave me Van Halen and Iron Maiden, and I walked back home and plonked them on the turntable. That was, I guess, Napster version 0.01. What Fanning did was take this idea into the twenty-first century, and the reverberations from it are still being felt.

In a *New Yorker* cartoon from 2006, a distraught-looking young man stands near the control panel of a crowded elevator, his hands near the glowing buttons representing the elevator's floors. Turning to his fellow passengers, he says, 'I'm sorry, I think I just pushed "shuffle".' This cartoon wouldn't have made much sense before the iPod's ascendance as the ubiquitous way to listen to portable music – not to mention, for a new generation, pretty much the *only* way to listen to music. It shows us just how prevalent Apple's incredibly influential music player has become.

iPods are wildly popular across all economic and racial boundaries (anything that is owned both by legions of teenagers *and* the Pope is a genuine cultural artifact). In the same way that Napster did, Apple's device changed the way millions listen to music. More interestingly, it altered how people buy and even *think* about music. Just as we now see the word *shuffle* in a new light, the word *download* has connotations it didn't have ten years ago. At the same time, these new zeitgeist terms are replacing phrases and words such as 'record store.' For the generation coming of age in the new millennium – let's call them Generation Download – all of this is normal.

While every group *besides* this new generation is talking endlessly about paradigm shifts, changing habits, and new memes for these kids it's not 'business as usual,' it's the only business they've known. They are the first generation to come of age not knowing anything but an existence with the Web. They can hardly imagine a time when the Internet did not exist or a world in which everyone didn't own a cell phone. These are kids who are wirelessly connected to each other at all times through text messaging, live blogging, or sharing videos and photos at social networking websites like MySpace and Facebook. It's an entire generation that has never known a classroom without a computer or a math class without a calculator, and can't remember a day when they weren't carrying around at least one electronic device at all times.

These aren't just gadgets, and these changes aren't superficial. An iPod hanging out of the pocket of a millennial youth is not the same as the slingshot that hung out of Dennis the Menace's red overalls fifty years ago. These new devices are epoch-defining inventions rapidly transforming the way an entire generation engages with entertainment. This goes far beyond the standard argument of digital versus analog.

'The Internet and digital media technology have become a way of life for most children and teenagers in developed countries,' wrote David Kusek and Gerd Leonhard in their 2005 book *The Future of Music: Manifesto for the Digital Music Revolution*, 'and mass media has become less relevant. This has created a behavioral shift that is one of the primary reasons for the huge popularity of file-sharing.'

Indeed, for this generation online connections truly are the 'extensions of man' that Marshall McLuhan wrote about forty years ago. Kids walk around with iPod headphones almost permanently stuck into their ears, while handheld videogames like Sony's PSP come with wrist-straps to make sure that portable entertainment is never far away. And of course, each of them has a cell phone. For them, the notions of space and time have been almost totally erased: all communication is instant and all information is just a mouseclick away. Even formerly physical objects, such as records and books – not to mention TV shows and movies – have been blasted apart and broken into minute digital slivers and chunks. And rather than try to put them back together again, Generation Download is sifting through the rubble, interacting with each piece of entertainment literally bit by bit.

One of the places we're seeing the biggest differences in the way Generation Download thinks and acts – as well as where we're witnessing the ripple effects this new behavior can have on society and economies – is in music. And what's changing in this new culture is not just the *style* of music, but the form (or format) of music itself.

'Every twenty years or so, a new generational tremor rips through the popular culture, signaling the arrival of new teens, and presaging an even greater pop earthquake to come,' wrote Neil Howe and William Strauss in their 2000 book *Millennials: The Next Great Generation*:

> Throughout the last century, every time a new generation has reached its teens, the sudden change in adolescent taste causes the engines of pop-culture production to stutter and stall. Then comes a period of trial and error, as the entertainment industry churns uncertainly until a new musical style catches on and thrives.

This change is happening again. This time around it's not merely one musical style that's triumphing over another (such as the way Punk was usurped by New Wave in the early 1980s). What we're witnessing is a new way of learning about, buying and listening to music. In fact, the recent rise of mash-up culture – combining elements from disparate sources in order to create something new, such as laying down vocals from Destiny's Child over the music of Nirvana – shows that, stylistically, Generation Download can't be pinned down in terms of genre or sound. The various music scenes are instead amalgams of sounds, attitudes, and styles; less hip-hop than hodge-podge. This has perhaps been best personified by producer Danger Mouse's now legendary pairing of *The White Album* by the Beatles with *The Black Album* by Jay-Z, creating from these two works a new listening experience, the download-only release *The Grey Album*. This 'record' – never available in stores, only on the Internet – was an instant sensation, and has been since downloaded well over a million times.

For Generation Download, the music they consume exists solely as the songs themselves. There's no package, no sleeve, no liner notes (except what appears online). Song titles appear only in the window of their iPods, and the music itself blasts through white earbud headpones without passing first through a pair of speakers. The days of kids going to a record store, flipping through new releases stacked in display cases, and then buying an actual record (be it on compact disc or whatever) are over. Much as The Buggles announced 'Video Killed the Radio Star' at the dawn of the MTV era, downloading music has changed the rules yet again. And the same way that the record industry was slow to react last time – at first hesitant about MTV, until they finally and wholeheartedly embraced it – the music industry in our new century has stumbled to find its way in the digital millennium.

Record stores, once a bastion of youth culture and a focal point for entire scenes (such as the Rough Trade record shop in the London, or Bleeker Bob's in Manhattan), now resemble homes for the aged rather than a youth hostel. A story in *The New York Times* from 2006 entitled 'The Graying of the Record Store,' discussed the generational shift now taking place on the frontline of the changing music culture. 'The neighborhood record store was once a clubhouse for teenagers, a place to escape parents, burn allowances and absorb the latest trends in fashion as well as music. But these days it is fast becoming a temple of nostalgia.'

Generation Download has no need to go to record stores. Software and websites bring the record stores to them. And with their headphones always on, and an electronic device in each hand, there's no need to leave the house in order to escape their parents; they can stare into their uneaten vegetables at the dinner table and still be in their own digital world.

The repercussions of this new behavior have been profound, the largest example being when the giant Tower Records chain went bankrupt in late 2006, laying off thousands of employees and closing hundreds of its record stores. While it had competition from more than just the Internet – huge retail outlets like Best Buy and Wal-Mart also cut into their sales – Tower had long discounted the threat of online music, and refused to change its business model accordingly. In fact, Tower founder Russ Solomon told a California newspaper in 2000 that the Internet 'is certainly never going to take the place of stores.' In an Associated Press story about Tower's demise, the president of Universal Music, Jim Urie, called Tower Records 'probably the greatest brand that will ever exist in music retail.' Now that 'great brand' is just a memory.

For kids today who bother with compact discs at all, it's usually only so they can burn a bunch of downloaded files onto a CD in order to give it to a friend. When this happens, any artwork that once accompanied the songs gets reduced to a magic marker scrawled across a silver surface for identification purposes only.

Talking about the difficulties the music industry is facing because of digital music, Alain Levy, the Chairman and Chief Executive of EMI Music, told an audience at the London Business School in 2006 that the CD 'as it is right now is dead.' Levy wasn't trying to be controversial. He was simply pointing out what has been apparent for the past couple of years: Generation Download, Millennials, Digital Natives – whatever you want to call them – are no longer buying CDs. The whole system for purchasing and consuming music has been completely changed by the iPod and the rise of digital music.

For an industry so used to selling a physical product, this has been incredibly hard to grasp. But Levy's forthright comment shows that some executives are finally beginning to face up to the fact that, whereas earlier generational shifts in musical consumption occurred from format to format (those raised on cassette versus those raised on eight track or vinyl), digital music has finally erased the need for a format at all.

All of this has revealed that Generation Download cares only about the content and the experience of the material; the songs themselves. This generation – which is already the most hustled and sought-after in terms of advertising dollars and disposable income – has discovered an important point: the music is all that matters. What this has meant to the music industry has been nothing short of monumental.

Digital music sales topped $1.1 billion in 2005, which accounted for 6% of the overall industry. That may seem like a small number, but to consider that just a few years ago digital sales were 0% of the industry, you can see in those figures just how quickly consumers adapt, and how quickly habits – many ingrained for years, some newly learned – can change. The fact that on 23 February 2006 iTunes celebrated its billionth download goes a long way towards showing that the old models are broken and that a revolution is well under way. And what was the name of the song that put iTunes over the billion mark? It was, appropriately enough, 'Speed of Sound' by Coldplay. Indeed, the speed of sound for Generation Download is instantaneous, traveling as it does over the Internet, and they will soon expect other forms of entertainment to follow suit.

All of which helps answer the inevitable question, Would anybody *really* read a book on a computer screen? Of course they would. If today someone willingly pulls up to a laptop computer or an iPod and listens to *Nevermind* as a series of digital files, and finds that it sounds just as fierce as it does coming out of a stereo playing the CD (not to mention without all of vinyl's snap, crackle and pops), so too will readers of the future get the same punch from *In Cold Blood* when they read it on some sort of electronic device rather than holding a paper version of it in their hand. The same also goes for movies. Box office receipts are down while ticket prices are up, and people would rather watch DVDs at home (or on their computers) than go to a theater. Also, small portable DVD players, in addition to handheld devices like Apple's iPhone show that consumers – of the Generation Download persuasion, especially – care most about the immediate access to the content.

In the same way that this new generation is eschewing traditional forms of media (not bothering to go the movies or watch television when shows are broadcast, and not buying CDs in stores), they will also be open to new methods of buying and consuming reading material. The very nature of Generation Download shows us that readers will one day (and sooner than we think) be more than willing to forgo an ink-on-paper book, and will not mind cozying up to their computer screens (or the screen of some device) instead of a physical book. We can see this coming because we have already seen how an entire generation has traded in their stereos for computers, not to mention an assortment of other portable electronic devices that keep them in constant touch with their digital worlds.

During a temporary Blackberry outage in 2007, Matt Richtel wrote an article in *The New York Times* entitled 'It Don't Mean a Thing if You Ain't Got That Ping.' The article talked about how (and why) Blackberry users felt so frustrated, stranded, and lonely without their Blackberries (after all, there's a reason that the devices are nicknamed 'Crackberries'). But the article also tied the Blackberry outage, and its feelings of withdrawal-like symptoms, to a more fundamental need of humans to stay connected in an increasingly electronic age:

> Experts who study computer use say the stated yearning to stay abreast of things may mask more visceral and powerful needs, as many self-aware users themselves will attest. Seductive, nearly inescapable needs. Some theorize that constant use becomes ritualistic physical behavior, even addiction, the absorption of nervous energy, like chomping gum.

Electronic content and networked books will undoubtedly help feed the 'visceral and powerful' needs which digitally connected people are often feeling. In fact, the quest for new stories that once drove readers to devour mountains of books now manifests itself in the young technophiles who feel the ardent tug to constantly be in contact via their electronic gadgets. But it doesn't stop there; this drive has spilled over into content itself. Users also want to interact with what they're reading, watching or listening to; they want to shuffle their playlists, remix their music, and alter how or when they watch movies and TV shows.

Richtel also discussed a condition known as 'acquired attention deficit disorder,' which is used 'to describe the condition of people who are accustomed to a constant stream of digital stimulation and feel bored in the absence of it.' I would say that pretty much anyone under the age of thirty qualifies for being accustomed to a 'constant stream of digital stimulation.' In our digital society, there's no escaping it. Ten years from now this will be true for nearly everyone. And so to expect future generations to be satisfied with printed books is like expecting the Blackberry users of today to start communicating by writing letters, stuffing envelopes and licking stamps.

In the early 1990s, critic and author Sven Birkerts was already running into this new phenomenon: students who were not only not interested in books, but seemed unable to muster the patience or interest to slog their way through even a Henry James short story. This encounter leads, for Birkerts, to a realization about this new generation, and its complete unwillingness – if not inability – to tackle the antiquated and wordy works of art that formed previous generations:

> But the implications, as I began to realize, were rather staggering, especially if one thinks of this not as a temporary generational disability, but rather as a permanent turn. If this were true of my twenty-five undergraduates, I reasoned, many of them from relatively advantaged backgrounds, then it was probably true for most of their generation. And not only theirs, but for the generations on either side of them as well. What this meant was not,

narrowly, that a large section of our population would not be able to enjoy certain works of literature, but that a much more serious situation was developing.

The NEA report *Reading at Risk*, issued nearly a decade after Birkerts began to first notice a change in his Generation X undergrads (who could almost be called well-read next to today's Generation Download), addressed similar – but growing – concerns. Indeed, the distressing patterns that Birkerts and other professors had begun to notice in the early 1990s, as a generation raised on MTV gradually gave way to one weaned on the World Wide Web, began to reach a depressing critical mass. And even though *Reading at Risk* outlines a host of declining habits, across a variety of socioeconomic levels, what many have found particularly distressing is what it has to say about Generation Download:

> Literature reading is fading as a meaningful activity, especially among younger people. If one believes that active and engaged readers lead richer intellectual lives than non-readers and that a well-read citizenry is essential to a vibrant democracy, the decline of literary reading calls for serious action.

The report also states that:

> the trends among younger adults warrant special concern, suggesting that – unless some effective solution is found – literary culture, and literacy in general, will continue to worsen. Indeed, at the current rate of loss, literary reading as a leisure activity will virtually disappear in half a century.

Marshall McLuhan sounded similar alarm bells in 1964, declaring that the battle was at our doorstep before most people realized it had even begun:

The electric technology is within the gates, and we are numb, deaf, blind, and mute about its encounter with the Gutenberg technology, on and through which the American way of life was formed.

But it doesn't have to be that way; Generation Download starts out with reading skills and levels comparable to recent generations. A 2006 study, funded in part by the children's publisher Scholastic, showed that reading habits in kids declined sharply after the age of 8, and continued to drop as kids became adolescents. According to a Scholastic press release, 'Almost half of the 15–17 year olds surveyed say they are low frequency readers.' So the generation who devoured Harry Potter has shown that they craved the content of those specific books rather than the act of sitting down with just any good, long book. Where they lose the love of reading is when they become teenagers, which is also when plugging into an iPod looks cooler than carrying around a paperback.

'Next to the new technologies, the scheme of things represented by print and the snail-paced linearity of the reading act looks stodgy and dull,' wrote Birkerts:

Many educators say that our students are less and less able to read, or analyze, or write with clarity and purpose. Who can blame the students? Everything they meet with in the world around them gives them the signal: That was then, and electronic communications are now.

This is how high the stakes are for book publishers. If they don't adapt to the habits of this new generation, they can forget about selling much of anything to them and those who follow. Yet, so far, most publishers are reacting cautiously if not indifferently, the same way that the music industry discounted the invention and rise of the MP3.

5

generation upload

WHEN I WAS IN high school in the mid-1980s, I put together a literary magazine that featured poetry, stories and art-work from my friends and other people at the school. One of the poems was my own, but I signed it using just my initials. It was a love poem titled 'To L.B.' This caused a minor scandal for about ten minutes as people tried to figure out who the writer was and who the subject was (it wasn't too hard). This and the stories I wrote for the school newspaper were the extent of my media exposure in high school. At the time, it felt like a lot. Whenever I saw someone reading our school news-paper I would get a chill down my spine, wondering if they were reading my words. On those day I walked the corridors of the high school feeling tall and brave.

But in terms of today's generation of Digital Natives, I would have been considered a hermit. Armed with inexpensive cam-corders, digital cameras and video cameras built into their cell phones, and combined with easy-to-use editing software that comes preloaded onto most laptops, Generation Download has seamlessly segued into Generation Upload. This goes well beyond the occasional MySpace or Facebook page; these are lives documented on a daily basis almost from the time they wake up to the time they go to bed, digitally preserved in the amber of pixels and mouse clicks. Remember that high school yearbook photo that you hate? Well, these kids will remember a lot more than that.

Not only are they sharing their own lives, but collectively they're creating huge online communities that harness the collected intelligences of the individuals into what's becoming derisively known as the 'hive mind.' The best example of this is Wikipedia, the user-generated encyclopedia where anyone can add or edit information about practically any or every topic that exists. Disparate, single voices which would have – even ten years ago – existed in solitude now join other voices in order to create a kind of online choir, where their multiplied presence adds up to something powerful and new. Now, suddenly, they belong.

This group of kids for whom the Internet has been omnipresent, and who have been hugely influenced in terms of their music and other entertainment choices by their ability to pull content down from the Web, now has the ability to post and upload their own creations. This allows them to instantly share their thoughts and experiences with dozens of friends (and millions of strangers). Computer screens are now two-way streets, allowing Generation Download to become delightfully schizophrenic by also giving them the ability to upload their various creations.

Of course, user-generated content is nothing new. From the Zapruder film to the videotape of Rodney King being beaten by the LAPD, ordinary citizens have long been capturing important events on film and videotape. These films or videos were then sent to news organizations to be broadcast or published, thus garnering extensive media attention. What the Internet has added to this process is the ability for users to easily put their own content onto the Web and then invite others to watch it, comment on it, and then pass it around to all of their friends. Distribution, once key in the dissemination of information, is now just a keyboard away.

In terms of news and documentary events, this is of course an amazing and useful way to show viewers footage from events shortly after they happen. During the bombing war between Israel and Lebanon in 2006, home video of bombs falling, and their aftermath – on both sides of the conflict – regularly showed up on video websites. While network news has been live ever since satellite technology has allowed reporters to embed themselves in hot spots, user-generated content has allowed a window into news happening in unexpected places, at unexpected times. When the Indonesian tsunami caught an entire nation off guard in 2004, the only visual

accounts of the disaster were from tourists with video cam
eras.

But none of this is entertainment. None of it is narrative
scripted or edited. Up until a few years ago, if someone
wanted to experience something like a sitcom online, a use
had to go to a file-sharing website to download an episode o
The Simpsons. But this was hardly revolutionary, since *The
Simpsons* was a TV show made for televisions; in this situation
a computer monitor was merely substituting for a TV screen
What would follow, in the wake of the rise of user-generated
content, would be narrative visual content made on comput
ers and cell phones in order to be consumed on computers
and cell phones.

But before there was an entire world of online video created
by amateurs, user-generated content was much more simple
What most people were creating and uploading were just
words, mostly in the form of blogs or online diaries. These
websites were places where people, pioneered by Generation
Download, would create blogs or simple websites where they
would talk about their lives, loves, and regularly post photo
graphs from their everyday life. All of a sudden the phrase 'my
life is an open book' was transformed into, 'my life is an open
website.'

In *The Gutenberg Galaxy: The Making of Typographic Man*,
Marshall McLuhan wrote about something which sounds very
much like the blogs and online diaries that gained popularity
at the turn of our century:

> In the electronic age which succeeds the typographic and
> mechanical era of the past five hundred years, we encoun-
> ter new shapes and structures of human interdependence
> and of expression which are 'oral' in form even when com-
> ponents of the situation may be non-verbal.

Blogs are indeed textual, and yet their breezy, conversational tone makes them seem like a friend whispering in your ear.

It's as if the audience – long since used to watching reality television shows – decided to make their own reality show by documenting their own lives and posting it on the Web for everyone to see; a truly *real* reality show. In Hitchcock's *Rear Window*, Jimmy Stewart's nurse calls the binoculars Stewart looks through all day, spying on his neighbors, a 'portable keyhole.' She means that he uses them to see into a variety of other people's lives. These new websites, blogs, live journals and online diaries are today's 'portable keyhole.' While the Web allows us to communicate with friends half a world away, it also lets us eavesdrop on the lives of strangers in other countries (or right next door).

What blogs have done is turned private thoughts and opinion into a global platform. In London, on Sundays, you can go to Speaker's Corner in Hyde Park and witness more than a dozen people of different political persuasions and social outlooks, each of them standing in front of sometimes huge crowds, bellowing out their opinions to anyone willing to listen. What the Web does, in theory, is allow anyone to reach thousands or millions of people, extending the range of that person's thoughts well above how far they could carry their own voice in a park or street corner. In *Annie Hall*, there's a cameo from Marshall McLuhan, whom Woody Allen produces in order to refute an obnoxious filmgoer standing behind him on line in a Manhattan movie theater. Today that obnoxious filmgoer would probably have a blog where he would pontificate daily about Fellini and Bergman, which people from around the world could read and comment on. Jerks have gone global; welcome to the age of user-generated content.

The Internet, through this capacity to easily reach millions of people, has already made 'stars' out of numerous ordinary citizens. Mostly unwitting participants have simply had compromising or else just plain embarrassing footage uploaded onto file-sharing websites. But, on the Internet, that's entertainment.

One of the most famous of these is Ghyslain Raza, better known as the 'Star Wars Kid.' He attained Internet infamy after a tape of him mimicking Darth Maul's lightsaber moves was put on a website by some schoolmates as a prank. Within a month of the video appearing, it had been downloaded over a million times. Numerous websites were erected around this two-minute video clip, and over a hundred variations of the video were produced (including a variety of mash-ups featuring the Star Wars Kid footage given numerous sound and special effects). Because of this, Raza was tormented everywhere he went in his small Quebec hometown. He sued, but eventually settled with, the kids who put the video on the Internet. The video continues to exist, and continues to haunt Raza; one website alone, dedicated to the clip, has recorded 76 million visits. But instances like these are mainly pranks, set up by a generation raised on *Punk'd*, for whom the Web is simply the new way to spread gossip or hurt someone's feelings.

With the advent of video websites such as YouTube, it has become even easier for people to upload and share videos. This has created a whole new world of 'clip culture,' where people create and share short video bursts of entertainment and self-expression. And this has not at all been confined to the Web, since the short bursts of content make them perfect to be viewed on smaller handheld devices like cell phones which is how some of the videos were filmed in the first place). But clip culture's real success has been YouTube itself, whose rise has been nothing short of a phenomenon. Founded in February 2005, YouTube has – in a few short years – become one of the most popular websites in the world. In October 2006, YouTube was bought by Google for $1.65 billion dollars.

The timing of YouTube's arrival was perfect, poised between Web 1.0 and Web 2.0, when users were hooked on the internet but wanted a more reactive and interactive experience. For a generation already bored by television, watching short video clips online wouldn't have seemed all that revolutionary. But when the invitation was extended to them to contribute their own clips, that's when things exploded.

'It's said that if you put a million monkeys at a million typewriters, eventually you will get the works of William Shakespeare,' wrote Bob Garfield in *Wired* magazine in 2006. 'When you put together a million humans, a million camcorders, and a million computers, what you get is YouTube.'

One of YouTube's first big stars was a user named 'funtwo' who uploaded a primitive video named simply 'guitar.' The video consisted of a non-descript youth, wearing a T-shirt and baseball cap, playing electric guitar along to a rockified arrangement of Pachelbel's wedding staple, Canon in D Major. The arrangement was by a Taiwanese guitarist named Jerry Chang, who had posted tablature for his arrangement on his own website, along with a backing track. When funtwo's rendition of 'Canon Rock' hit YouTube, it was a smash hit. All of a sudden the video's creator, 23 year-old Jeong-Hyun Lim, was an Internet celebrity garnering the kind of attention – and numbers – that major label acts would kill for.

'If individual viewings were shipped records, "guitar" would have gone gold almost instantly,' wrote Virginia Heffernan in *The New York Times*. 'Now, with nearly 7.35 million views –

and a spot in the site's 10 most-viewed videos of all time funtwo's performance would be platinum many times over.

But while the 'guitar' video was almost purposefully vague (funtwo wasn't out to achieve fame; in fact, it was a while before Lim stepped forward as the creator of the clip, which in the meantime gave rise to numerous impostors), there are others who are using YouTube precisely because of the exposure it can bring. These new 'online auteurs,' as described by *The New York Times* in 2006, are using the Internet in order to launch their careers. For them, the Internet provides a crucial platform to create and reach an audience, and thus gain exposure that will lead (they hope) to mainstream riches and fame. 'Who needs film festivals?' the *Times* asks. 'YouTube videos are viewed 100 million times a day.'

In the process, novice filmmakers and comedians have used the Web to create short films that have brought network-size audiences and created new stars, like lonelygirl15, Chad Vader, Ze Frank, and the ninja from, well, *Ask a Ninja*. All of these shows have logged million of views and downloads, and each were created by industry outsiders for very little money.

Between this and the success that Danger Mouse had with his mash-up *Grey Album*, the audiences that these homemade projects reach are enormous and deliver numbers that would impress the major studios or record labels that didn't think to produce them in the first place. Indeed, what gives these online shows so much of their appeal, at least from the point of view of those watching them, is that they *aren't* affiliated with a television network or other professional company or studio. For a generation so used to – and burned out on – corporate placement and sponsorship, the idea that these creations come from individuals just like themselves is intoxicating. And, for the creators, the ability to reach a mass audience so cheaply and quickly is also intoxicating. As described in the *Times* in 2006, 'Now you can shoot a movie on your cell phone, transfer it to your computer and post it on YouTube three minutes later. It's instantaneous, it's free, it's idiot proof.'

Even if the audience is party to a scam, such as lonelygirl15 (many viewers knew, after only a few 'webisodes,' that something was amiss and that this wasn't a real girl who was 15 and lonely), they often don't care. What then happens is that the drama turns into a mystery, and the sheer act of viewing the clip leads them to begin to wonder about the origin of the clips. They know it's a fake, but they then begin to have fun trying to figure out how fake it is.

'Many assumed the series would sputter and die,' wrote Joshua Davis in *Wired* magazine late in 2006, after it had been revealed that *Lonelygirl15*'s 'Bree' was actually Jessica Rose, a 19-year-old actress. 'Media reports zeroed in on how viewers had been duped, suggesting an inevitable backlash. But the fans – raised on the unreality of reality TV and with the role-playing ethos of the Web – seemed to take the revelation in stride.'

YouTube and other video outlets are creating more than just their own homegrown stars. Many different kinds of performer are harnessing the power of YouTube and similar sites in order to reach out to their existing audience and also to get new ones, gleefully bypassing the mainstream media as they do so. For instance, many bands are making their own videos – or having their fans make videos for them – which are then posted on video websites. By doing this, these bands have saved tens or hundreds of thousands of dollars by not having to create slick, polished videos (not to mention they don't have to try to compete for the limited amount of air time that music videos now receive on MTV).

The best example of success in this area has been the Chicago-based band OK Go, who created a video for their song 'Here it Goes Again.' The video was shot with one camera angle, and featured nothing but the four band members in a room with six treadmills. As the music plays, each member jumps from treadmill to treadmill with the grace and ease of Fred Astaire. It doesn't sound like much of a premise, but the end result is campy, classic and hypnotizing. The band made the video on their own, and fans from all over posted it on YouTube (where it has since been viewed millions of times). This exposure led to increased notoriety for the band, including appearances on news shows (mainly to explain the popularity of the video), as well as award shows (such as the American Video Music Awards, where they performed their treadmill dance live).

'Music industry watchers can learn from OK Go's experience,' stated Reuters, 'which shows that Web users can catapult a band to fame, challenging the popular assumption that videos need to cost thousands of dollars or be directed by Hollywood filmmakers.'

What big-time video directors like Spike Jonze used to create artificially in order to comment on and criticize slick, MTV production values – such as his purposefully amateurish Fatboy Slim video for 'Praise You' – is now being accomplished by real-life people and indie bands, doing and paying for it themselves.

For OK Go, this was just the beginning. The 'treadmill dance' video, as it quickly became known, ended up copied, parodied and reenacted dozens of times in amateur videos (which were also uploaded to YouTube and various other websites). This led to even more exposure for the band. But what this has also done, perhaps more importantly, is that it's given the fans a way to participate in the process. As far back as 1980, Alvin Toffler dubbed them 'prosumers' in his book *The Third Wave*. The term is a mash-up of the words 'producer' and 'consumer,' signifying that the lines that once separated the two have been erased.

n the 1970s, as a young kid living in the California suburbs, I vas a huge fan of the rock band Kiss. While always popular, hey never had any mainstream chart success. Instead, Kiss ictively cultivated a rabid following of fans who flocked to heir shows and bought their records. Labeled the 'Kiss Army,' these legions of fans would apply make-up to their aces before the shows, and knew all the words to Kiss songs by heart. Every Kiss record inevitably came with an insert which sold paraphernalia to Kiss Army acolytes: t-shirts, headbands, patches, buttons, jackets etc. The army had a uniform, but that was about it. The Kiss Army was all about consumerism and buying products; being, well, a member in an army and just another soldier blending into the crowd. (I myself was a proud member of Kiss Army, and despite this I never got closer to Kiss than an album cover.) But today, the internet is actually allowing interaction with musicians; fans aren't just part of the act, they're crucial to the act being here in the first place.

'This is not merely an illusion of intimacy,' wrote Clive Thompson, describing what he called 'Artist 2.0' in a 2007 *New York Times Magazine* story:

Performing artists these days, particularly new or struggling musicians, are increasingly eager, even desperate, to master the new social rules of Internet fame. They know many young fans aren't hearing about bands from MTV or magazines anymore; fame can come instead through viral word-of-mouth, when a friend forwards a Web-site address, swaps an MP3, e-mails a link to a fan blog or posts a cellphone concert video on YouTube.

Gone is the recluse or moody dilettante who has no interac
tion with his or her fans; in a digital world, musicians rely on
their fans to not only buy their work but also to offer advice
cover their songs, make their videos, and help them book their
tours and be a part of their live shows.

New York hip-hop legends The Beastie Boys embraced this
concept when they handed out fifty video cameras to fans
before a sold-out show at Madison Square Garden in 2004
asking ordinary people to film the concert from their own
unique perspectives. Beastie Boy Adam Yauch then used all of
the footage to create the almost entirely fan-made feature
film, *Awesome: I Fuckin' Shot That!* Yauch told *Wired* magazine
that he was inspired to have fans create the movie after seeing
a video clip posted on the band's website:

> It was really grainy and shaky, but I loved how it was shot
> from eye level and showed a personal take on what was
> happening onstage. I thought it would be cool to apply
> that approach to a full-length concert film, so that it actu-
> ally feels like you're watching a concert and not some big,
> overblown MTV video.

Musicians are not only turning to fans for inspiration, but
are even letting them contribute to the form of the final
product. Eric Steuer, commenting on Beck's record *Guero*
which was released in a variety of editions and formats in
2004 and 2005, wrote in *Wired* magazine that there really is
no definitive thing as *Guero* 'because there was no album, no
static list of 13 songs... Such is the future of the album, as
envisioned by Beck; it's something to be heard, seen, recon
stituted by artist and audience alike.' The listener can either
choose which version of the record they'd like to experience
or else mash up their own version from any of the existing
versions, or else – using the individual tracks for certain songs

that Beck posted to his website – create their own mixes and versions of Beck's songs.

Beck's next record, 2006's *The Information*, went even farther, allowing the owner of the record to listen to the music on a CD, watch videos for every song on an accompanying DVD, as well as design their own cover using a packet of stickers that also came with the package. *Wired* called it the 'Infinite Album,' and with it Beck set out to challenge the way both his fans and the music industry look at records. In terms of the music industry, he did indeed shake things up, to the point where *The Information*, in the United Kingdom at least, was deemed ineligible for chart status since the UK's Official Chart Company deemed that the stickers and packaging gave the album an 'unfair advantage' over other releases. It seems that, so far anyway, the industry is not yet ready for these kinds of changes. But what's important is that the fans are indeed ready, and expect more artists to follow suit in letting them participate in some way in the listening experience. The repercussions all of this could have for the music industry, already reeling from the decline in CD sales and the rise of digital downloads, are huge.

'The very logistics and economics of the music industry are at stake,' wrote Eric Steuer in the same Beck article from *Wired* magazine, 'as one album becomes a long shelf of songs and products, each carrying its own release date, distribution path, and price tag. In the end, fans can create their own versions of the album, stringing fave songs and remixes into one ideal playlist.'

Fans have been doing this for years by themselves, creating mix tapes or else simply with their stereo remote control, skipping songs from across the room or deleting that noisy first track or ending a CD prematurely instead of wading through the four (or forty) minutes of silence at the end to get to the brief 'hidden' track. What had been previously difficult with vinyl – having to get up, cross the room, pick up the needle and place it back down on the record at the appropriate blank space between the grooves – in the CD age was incredibly simple. That is, if you had the time to sit there with the remote and manually create the record as you wanted to hear it each time you heard it. Most multiple CD players allowed users to program in primitive playlists, shuffling discs back and forth and cherrypicking songs from each, but the delay between songs as the discs replaced each other made this not very useful for longer sessions.

But the new technology of software like iTunes allows users to automate this kind of behavior, creating once and for all the record *they* want to hear. For example, it has been a long-standing criticism that most double records are sprawling, messy affairs that would usually create one great disc (The Clash's sprawling, three-disc *Sandinista!* could do with some judicious pruning.) Users can now create their own versions of any album by creating, as Beck's recent projects have hinted at, an album of their own 'infinite' choices and possibilities.

Another area of user-generated content that began as a way for Digital Natives to interact with their entertainment, but which is now becoming a bona fide art form, is machinima. The word is a combination of 'machine' and 'cinema,' and is used to describe feature films or short clips that are created when gamers manipulate the characters in video games in ways divorced from the action, recording their movements on video and then writing dialogue for them to say and dubbing it on later. This can lead to hilarious results.

The best known machinima creation is *Red vs. Blue*. This is a series of films created by a group of gamers in Texas who use the violent war game *Halo* in order to create short narrative clips of two opposing soldiers (one red and one blue, hence the name), standing around on the battlefield talking about absurd or mundane topics; it's kind of like *Platoon* meets *Seinfeld*. *Red vs. Blue* has been an instant Internet sensation, with millions of people downloading the free episodes as soon as they're made available, and offering to buy DVDs of the collected episodes.

Machinima allows gamers everywhere to cheaply and quickly produce amazing computer animation, and then to make up dialogue to put on top of the images, usually subverting the original subject matter in much the same way Woody Allen and his cohorts did with *What's Up, Tiger Lilly?* back in the 1960s. The allure, according to the machinima.com website, is that 'you can produce films on your own, or you can hook up with a bunch of friends to act out your scripts live over a network. And once you're done, you can upload the films to [the machinima site] and a potential audience of millions.'

Of course, friends getting together to make movies is nothing new. The legendary B-movie director Ed Wood, in the 1950s, basically did the same thing well outside the Hollywood mainstream. The difference with the online auteurs of the early 2000s is that the Internet now offers distribution as well. Wood and many others like him, once they had made their movies, were then dependent on movie theater owners across the country to show their films; it was no use making a masterpiece if no one could see it.

Even *Citizen Kane*, often regarded as the greatest movie ever made, received scant distribution at the time of its initial release. Theater owners everywhere boycotted the film due to threats from William Randolph Hearst, the thinly veiled subject of the film. Orson Welles was so frustrated by not getting *Kane* shown in theaters that he had the idea to drive around the country showing the film in tents he would set up expressly for the purpose. The movie never did receive the acclaim it should have in its day, and the lackluster reception it received doomed Welles to a tragic life led at the margins of the film business. In his later years he worked mainly in Europe, and had to take bit parts, commercial work and voiceover jobs in order to fund his films, many of which – including *Don Quixote*, *The Other Side of the Wind* – he left uncompleted at his death in 1985.

Internet filmmakers today now have an immediate outlet for their work, with millions of people looking to the Web and to popular video websites for entertainment. The true underground now flies through the air in wireless connections, and the ability to reach millions of people – almost instantly – is at the fingertips of just about everyone.

Kids who have spent hour after hour playing games like *Halo* and *World of Warcraft* – which in their gameplay and storylines are already incredibly immersive – have shown that they want more from a game than to just *play* it. They want to become part of the game, interact with it, and have an influence on the action and the game's surroundings. Passivity is out; interaction is in. Who even knew what Pac-Man was twenty-five years ago, let alone wanted to interact with it beyond moving him around a maze eating power pellets? This new generation, more than any before, is part of the entertainment it consumes.

The implication for books and literature is that a new generation of kids, weaned on being 'prosumers,' will want to interact with and, to a degree, create the material that they read. 'Because [the Net generation] are used to highly flexible, custom environments which they can influence,' wrote Don Tapscott in his 2006 book *Growing Up Digital*, 'they want highly customized services and products. They are as used to having options as they are to breathing oxygen.'

In a *New York* magazine cover story from 2007 entitled 'Say Anything,' there was a list of the various changes between Generation Upload and previous ones. The first item was 'THEY THINK OF THEMSELVES AS HAVING AN AUDIENCE.' This is radically different from previous generations, who considered themselves just *an* audience. In terms of publishing, today's kids are not going to want to pick up a big book and spend hours in a corner silently, passively reading. Why in the world would they do that? It's not interactive. They can't share the experience with their friends. There's no way to change the book to suit their own tastes. Instead, they're going to ditch the hardback and head over to Facebook.

The publishing industry needs to realize this, and it needs to also find a way to get to these kids by making content available in a way that will first reach them (i.e. digitally) and then will give them the tools to interact with it and share it (post excerpts on their MySpace pages, email chapters to friends, IM paragraphs across class etc.). If not, there are dozens of ways this generation will choose to spend their time, and none of them will involve books.

The syndication of news and other content-heavy websites is the first nod in this direction, allowing users to receive only the portion of a publication that they're interested in. Why would someone want to receive the content of an entire newspaper when all they're interested in is sports, or the front page? Similarly, books as we have known them for hundreds of years – static, unchanging, silent – will have to change, perhaps in such a way that they're shattered to allow for user manipulation. Non-fiction texts, especially, will be broken down into bits and pieces, 'microchunked' for the consumption of individual parts, consumers picking and choosing chapters and passages from different books as if they were at a buffet.

Of course there are many who contend that books are works of art and shouldn't be reworked or touched at all. The latter is of course a silly view since readers 'rework' books all of the time by skipping whole sections as they read, the same way that people rarely ever listen to the entirety of *The White Album*. We each make our own personalized version of an art form whenever we choose to experience it. I may rent a movie and watch only half of it, or else listen to only eight out of ten songs on a CD, or else dip in and out of a novel as I read it, skipping entire sections along the way. This is the way it's always been; consumers are in charge of something once they get it into their hands, and it's ridiculous to think that they're not.

Once books become more widely available electronically it'll just be a matter of time before a generation raised on the user-generated content of YouTube, mash-ups and machinima starts to interact similarly with its text. Whether that means cutting out the boring bits of *The Mill on the Floss*, or else remixing *Middlemarch* and *Middlesex* until the hermaphroditic saga of Eugenides is transported to the 19th century world of Eliot, remains to be seen. But there's going to be no stopping upcoming generations from mixing and matching – and then sharing – the words that they read, and writers and publishers need to start to get comfortable with that fact (not to mention that they should acknowledge that this will be a positive development). Or would they prefer that future generations not read this material at all?

In her 1998 book, *How Reading Changed My Life*, novelist Anna Quindlen predicted an age where texts could be modified by users in much the same way that kids today create their own versions of records by turning them into playlists. Quindlen imagined reading 'The Fountainhead via the Internet, perhaps with all the tiresome objectivist polemical speeches set in a different font for easy skipping-over (or even the outright deletions that Ayn Rand's editor should have taken care of).' The ability to alter, and then share, text to this degree would mean that you could edit a book to your own liking and then send an amazing chapter or even a couple of sentences to someone, via email or a webpage, along with a message that says, 'Take a look at this; I think it's amazing.' Imagine all of the sharing of literary material that would occur if the reins were loosened just a little. The kinds of textual samplers that kids would share and email back and forth could be like mix tapes of the 1980s, where friends put together selections of music, giving them to other friends in order to impress or woo them. Or people could, as Quindlen predicts, prune books to their own liking.

In the face of Generation Download's changing taste, the music industry did not take such a pragmatic view of things. For a long time, record companies fought against the breaking apart of their content. 'From the music industry's perspective, of course,' wrote Steven Levy in his 2006 book *The Perfect Thing: How the iPod Shuffles Commerce, Culture, and Coolness*, 'keeping the package whole wasn't an artistic consideration but a commercial one.'

The record labels finally relented, and Apple proceeded to change the way a whole new generation listens to music: one song at time. Consumers now expect some form of this. And while this may seem jarring to lifelong readers used to sitting down with a hardcover book, starting on page one and flipping through the pages consecutively until they get to the end, this way of experiencing art – or just about anything – now seems hopelessly old fashioned. 'For this younger audience,' writes Douglas Rushkoff in *Playing the Future: What We Can Learn from Digital Kids*, 'discontinuous media is not the exception, it is the rule.'

There have been many contributing factors that have led to this point, but one of these is the iPod, which has shown just how into non-linear experiences Generation Download is.

'From following the iPod since its inception, both as a reporter and someone bound to his subject literally by the ears,' wrote Steven Levy, 'I came to understand that one feature in particular was not only central to the enjoyment of this ingenious device but has come to symbolize its impact on the larger media landscape – and perhaps to embody the direction of the digital revolution in general. Shuffle.'

Whereas Generation Download defined themselves by pulling material from the Internet, discovering and experiencing content in a new way that was revolutionary because of its quickness and formatless delivery, Generation Upload is beginning to define itself by mixing, mashing, and combining disparate elements of what they've pulled from the Internet, and then changing it into something else. These new, user-generated works are then uploaded to websites for the entire world to see.

This is a new generation wanting to interact with its music and become part of the games that they play; they are not content to sit and be passive. Publishing is going to have to, on some level, allow for the customizing of their texts or material. If not, Generation Upload will forget about books that, in hands eager to have an impact on everything they touch, will seem stubborn and unyielding. Instead, they will simply turn to something else, something that they *can* interact with.

6

on demand everything

UNTIL VERY recently, television networks scheduled shows on a specific night of the week at a certain time of day, and it was then up to the viewers to be sitting in front of their television sets at that time or else they'd miss the show. For decades now networks have worked like this, with network executives spending days and nights in large conference rooms sweating and debating over the various lineups for the upcoming season, the names of shows on little bits of plastic shuffled around the days of the week like checkers on a checker board. What should they put on Mondays at 9:00 p.m.? Or Saturdays at 10:00 p.m.? What's the competition putting on at the same time? They're hoping to hit gold with not just one show at one time, but to corner the market on an entire time-slot and, if possible, an entire night.

In the 1990s, NBC built up a huge audience on Thursday nights with their 'Must See TV' line-up. This included a clutch of their most popular shows, such as *Friends*, *Seinfeld* and *ER*. And people all over America did indeed stay home in order to watch these shows, plunking themselves down in front of their television sets week after week, afraid to move a muscle from 8:00 p.m. until 11 p.m. lest they miss something they 'must' see.

This is how the television industry has operated pretty much since the invention of the medium, making the business model of television clean and simple: the more popular the show, the more the network can charge advertisers to show commercials during the breaks during and between those shows. People of course put up with the commercials because they want to watch the shows more than they don't want to watch the commercials. But now, that's all changed.

Technology stepped in and allowed people to watch the shows they want to watch independent of the timeslots the network originally slates. People are now freed from their couches, able to live their lives and not 'miss' anything. If plans come up on a Thursday night, you can go out to dinner and see friends, come home at midnight and still see *Friends*.

Since the early 1980s VCRs had allowed viewers to do pretty much the same thing, but recording multiple shows on multiple nights – on videotape – was always a clumsy proposition. The machines themselves were never very intuitive, and people didn't want to go through the hassle of buying the tapes, setting up the tapes, labeling them, taking one tape out and replacing it with another etc. Wrangling with all that videotape wasn't very liberating; instead of being chained to the couch, viewers were tied up with tape. So even after the advent and ubiquity of videocassette recorders, most people were still sitting in front of their televisions, watching shows when the networks – in their infinite wisdom – decided to put them on.

Then along came digital video recording (also know as DVR, which goes by a few brand names: TiVo, DirectTV etc.). Viewers now can record their favorite shows (the entire series, or else individual episodes), rewind and pause live TV, watch two shows at once, record two shows at once, or even record two live shows while playing back an episode of something recorded last week.

While networks still program certain shows at certain times, hoping to gain audiences at specific points in the day, DVR technology has now given the power to the viewers, who can watch those shows whenever they want. In the age of digital video recording (in addition to video streaming, where people can watch their favorite shows on the Web), 'prime time' is whenever people 'find time' to watch whatever it is they've recorded or searched for online. 'Empowered by the Internet and digital video recorders, massive numbers are resetting their TV clocks,' wrote Johnnie L. Roberts in *Newsweek* in 2006. 'Viewership is beginning to spike during daylight hours, when videos are streamed or downloaded to office computers, laptops, iPods and cell phones.'

The ripple effects from these changes are being felt at all levels of the television industry, from the studios producing the shows to the advertisers creating and placing the ads during the commercial breaks of those shows. Everything is up in the air, which means that everything is also up for grabs.

'All the old definitions of TV are in shambles,' wrote Jeff Jarvis in *The Guardian* in 2006:

> Television need not be broadcast. It needn't be produced by studios and networks. It no longer depends on big numbers and blockbusters. It doesn't have to fit 30- and 60-minute molds. It isn't scheduled. It isn't mass. The limits of television – of distribution, of tools, of economics, of scarcity – are gone.

For advertisers, this is upsetting the delicate balance of a symbiotic relationship that had flourished for decades. Faced with the commercial breaks in digitally recorded programs, consumers just fast-forward right through them.

Just as Generation Download changed the face of music by altering the distribution and consumption of songs and albums, and Generation Upload has created a new array of Internet stars (many of whom disdain the mainstream media and want only to build upon their online success), so too has the on-demand model similarly changed the face of television. Chronicling the rise of YouTube for *Wired* in 2006, Bob Garfield wrote that 'fragmentation has decimated audiences, viewers who do watch are skipping commercials, advertisers are therefore fleeing, the revenue for underwriting new content is therefore flatlining, program quality is therefore suffering.' In many instances, the rise of digital media in other formats, and the Internet in general, has already led younger generations away from the television set.

For the latchkey kids of the 1970s the television may have been their best friend, 'but to today's media-literate kids, television's current methods are old-fashioned and clumsy,' wrotes Don Tapscott in his 2006 book *Growing Up Digital.* 'It is unidirectional, with the choice of programming and content resting in the hands of few, and its product often dumbed-down to the lowest common denominator.' The bottom line is that not as many kids watch TV as they used to, and those who still watch television want to watch it in a way far different than previous generations sat in front of their TV sets.

All of this goes far beyond network television and the notions of 'prime time.' The changing habits of consumers who want to experience their entertainment anytime, anywhere, have also begun to reach into the movie business.

Before DVD players made them extinct, the way to watch movies at home was on a VCR. People would rent movies from their local video store (before Blockbuster put it out of business; in the 1990s, Blockbuster Videos were to mom-and-pop video stores what Starbucks are to mom-and-pop coffee houses today). Consumers would then try to find the time to watch the movie in the allotted rental period – usually two or three days – always living in fear of returning the tape late and thus paying a fee (or else paying extra for not rewinding the tape; a concept that would be lost on Generation Download, for whom the idea of something that stores information in a linear fashion – with a beginning, middle and end – is alien). Now people get DVDs in the mail from Netflix, which they can then hang on to for as long as they want. Or else they can dial up a specific movie from an on-demand channel, 'renting' access to it for a couple of days with the ability to then watch it whenever they have time. The charge for the film then shows up on their cable bill. Or else they can buy and download individual movies from any number of websites, iTunes included. They can then watch the films on laptops, handheld devices or video iPods.

In *The Player*, Robert Altman's 1992 satire of the movie business, a bunch of executives sit around in a conference room on a movie lot, trying to think of ways to decrease the ballooning budgets that are ruining their business. One of them zeros in on the high cost of hiring screenwriters. If only they could get rid of screenwriters and come up with the stories themselves; they'd save millions.

It of course seems ludicrous; how can you have movies without a screenwriter? And yet YouTube and clip-culture (not to mention reality television) have proven that you don't need scripts or screenwriters. And now, our modern on-demand everything culture is proving that you can have movies without movie theaters. Of course, the easy way to do this is simply to shun going to the movies and wait for the film to come to cable or DVD. But a new idea is growing that will allow potential viewers to experience a movie in a variety of formats when a film is initially released, all at the same time.

The first to really embrace this idea was Steven Soderbergh with his low-budget 2006 film *Bubble*, which was released in theaters, on DVD and on high-definition cable TV, all on the same day. For viewers, there was no waiting game in terms of choosing between the theater or the couch. 'I don't care how people see the movie,' Soderbergh said at the time of the film's release, 'as long as they see it.' And more directors and studios are beginning to feel the same, realizing that by giving the consumer a choice between formats – simultaneously – they are *increasing* their chances of gaining an audience and making money, rather than *decreasing* them. More choices simply mean more chances for people to pay for their product.

Soderbergh predicts that within five years every big Hollywood movie will go out in all formats at once in order to capture the audience's attention (wherever it may be). If someone wants to watch a movie in a theater with surround sound, or watch it on a high-definition television at home (or else on a laptop on a train during a commute or a video iPod while sitting in the middle of Central Park); in any of these situations, the material must be available. The hard part, as Soderbergh, points out, isn't how they watch it but getting them to want to watch it in the first place.

Not every director feels this way. In a 2006 interview with *The Hollywood Reporter*, Steven Spielberg heaped disdain upon the idea of watching a movie on an iPod's 3-inch screen. 'That's one medium,' he said, 'where I have to draw the line.' But in today's digital age it's no longer up to a director how his movies are experienced or enjoyed. The consumers, with all of the choices available to them, now have the final word.

Spielberg may not direct a movie keeping 3-inch screens in mind, but consumers may choose to watch his movies on such a device. Of course, if he or his studio fails to release his films in the digital formats that can be legally downloaded onto handheld devices, then they can expect hackers to make them available (probably for free). In fact, when asked about his decision to release *Bubble* in all formats at once, Soderbergh replied that simultaneous release already exists in abundance. 'It's called piracy,' he quipped.

Spielberg is also quoted in the *Reporter* as saying, 'I don't think movie theaters will ever go away.' And he's right; movie theaters will not become extinct any more than bookstores will become ghost towns. However, in a world of on-demand everything where consumers can enjoy just about any kind of entertainment at home, the need for people to go to movie theaters will decrease significantly for a variety of reasons: it's very expensive to go to the movies, it is quite often not an enjoyable experience due to the large and loud crowds, and people have shown they don't mind waiting three or four months for the DVD or the film to appear on cable. Also, there are just more things for them to do *besides* watch a movie.

There's a scene in an early Alfred Hitchcock film (*Rich and Strange*, from 1931), in which a restless married couple is looking for a way to spend their evening. When the husband comes home from another long day at the office, his wife asks if they should 'stay in and listen to the wireless or go out to the pictures.' Those may have been the only two choices for entertainment that existed seventy-five years ago, but for a couple today who have at their fingertips TiVo, Netflix, iPods, satellite radio and the Internet, which acts as a portal to just about any form of entertainment that exists (music, film, literature, television) – making all of this material available immediately – the opportunities for entertainment have exploded. The days of having only one or two options are long gone. This is what all makers of entertainment are up against in the digital age. We cannot wait for consumers to adapt to formats; formats must adapt to consumers.

The battle is no longer between timeslots or TV shows, it is for attention: not the shows that potential viewers watch, but to try and get them to watch TV at all.

'It's easy to start a business, to get access to customers and markets, to develop a strategy, to put up a Web site, to design ads and commercials,' wrote Thomas H. Davenport and John C. Beck in their 2001 book *The Attention Economy: Understanding the New Currency of Business*. 'What's in short supply is human attention. Telecommunication bandwidth is not a problem, but human bandwith is.'

In view of the way entertainment has been seen as a commodity for the last hundred years – available on the terms of big studios or music labels – you can see how revolutionary this sounds. Fans of music used to be tied to their stereos, but along came the iPod. This new gadget allows them to take their entire music library with them – tens of thousands of songs – so that they can listen to anything wherever and whenever they want. Television has been similarly set free (if not yet so dramatically in terms of portability, though that will happen soon). And now that films are released so quickly onto DVD, not to mention that certain studios and directors are experimenting with a one-day release schedule, film fans can increasingly watch new movies on their own terms rather than only when their local theater happens to schedule a screening. Will publishing follow suit?

Because while we would hope that everyone has the time to read uninterrupted for three hours a day beside dimly lit fireplaces in cozy dens, the reality is that most people grab reading time whenever and wherever they can. In big cities this means during a commute: magazines, novels and newspapers read on subways, buses and trains. Or else people will read while they do their laundry, wait in a doctor's office, or during a lunch hour while they're eating. I mentioned in the first chapter that one of the strikes always held against eBooks is that they can't be read in the tub, but who nowadays has the time to light candles and lounge around in a bubble bath reading a fat novel? Doesn't everyone take quick showers?

In a 2007 *New York* magazine article by Ira Boudway entitled 'The Media Diaries,' three New Yorkers 'track everything they watch, read, and listen to in the course of a week.' What I found interesting is how digital their lives are, especially a woman who bought an episode of the television show *Grey's Anatomy* on iTunes and watched it on her laptop. All of them read news online, in addition to dozens of other websites. Also, there aren't many diary entries like '1 p.m.–6 p.m., sat in a bay window and read Tolstoy.' True, these are New Yorkers, and so may not be representative of the rest of the country, but I think it's not too far off the mark of how lots of young people are now living.

'When *Time* magazine put a crinkly, vaguely toxic-looking fun-house mirror on its cover and named "you" the person of the year for 2006,' wrote Boudway, 'the Establishment weekly was more or less cheering on its own diminution. After all, like most purveyors of mass media, from TV (see the nightly news) to the music industry (Tower Records, R.I.P.) to daily newspapers (which have lost over 20 percent of their stock valuation in the past four years), *Time* is facing both a vexing shift in consumer behavior and the rise of self-generated content. Of course, amid all this apocalyptic hype, young people are consuming more media than ever.'

The real change is occurring in *where* and *how* young people are consuming their media; more often than not, media are being consumed outside of their natural habitats. TV shows are not watched as they're broadcast. Records are listened to as electronic files while their CDs lay in their jewel cases.

While ordering something on Amazon and paying for overnight shipping is *almost* an on-demand environment, it's not the same thing as getting a recommendation on a band, heading to iTunes, and downloading three of their songs with three clicks of the mouse. In an online world, where studies have shown that 75% of users shopping on the Internet won't return to a website if it takes more than four seconds to load, twenty-four hours is an eternity. Today's shoppers want it *now*. And *now* is a lot faster than it used to be.

Prose will be left behind unless it makes strident efforts to adapt to this 'I want it now' on-demand model. Electronic versions of books or text – instantly downloadable or accessible upon purchase – will meet the expectations of generations used to having everything else available on demand, especially when the material can be accessed at any time through an online bookshelf.

While this will lead to more people reading (since they'll be able to get access to their reading material wherever they are), it will also lead to more discovery. If a consumer is searching for an author or topic on Google (say, 'Truman Capote' after seeing the film *Capote)*, if they see in a search result that there's a book about or written by the subject, they'll be able to purchase the text with a credit card, downloading it to their device or computer, or else they'll be able to buy instant online access to the material or even rent the text for a specific period of time. Whatever the ultimate business model behind the purchase, within seconds they'll be off and reading. What used to take a drive to the bookstore, or a trip to the mall, or even – just five years ago – a few mouseclicks and then a few days' wait from Amazon, will now take barely a few minutes if not seconds.

There will also be an impressive market for smaller parts of entire books. The same consumers who buy one song at a time on iTunes, or else one episode of a TV show, may want to purchase just a chapter or even a few pages of certain kinds of books. For some genres it makes perfect sense to provide excerpts: travel guides, cookbooks and certain kinds of reference and scholarly material. Why buy an entire guide to Italy when you're only going to visit Sicily? While this would be rarer for novels – after all, no one's clamoring to buy half a movie – the rise of YouTube and 'clip culture' has shown that consumers are increasingly looking for non-linear, bite-sized bits of entertainments. 'Snippet culture' may not be far behind.

Of course, the numerous pundits who say that readers will never want to part with paper will also add to the argument that no one will ever want to divide up a book like a grapefruit, buying and consuming selected sections instead of the whole thing. Yet a similar shift has already occurred in the photography business, with digital cameras overtaking film cameras. Consumers, it turned out, wanted to be able to pick and choose from their numerous shots, rather than getting whole rolls of film developed. Because why have prints of all thirty-six photos when you really only want three or four?

So what happened in terms of the disappearance of the CD, the exploding of TV, and the shrinking numbers of movie-goers, will also happen to books; the more our society turns to technology, the more things technology will touch. Newspapers are already providing customized content in the form of RSS feeds; trade publishing will have to follow suit.

'Spurred by Google's initiative and by the lower costs, higher profits, and immense reach of unmediated digital distribution,' wrote Jason Epstein in *The New York Review of Books* in 2006:

book publishers and other copyright holders must at last overcome their historic inertia and agree, like music publishers, to market their proprietary titles in digital form either to be read on line or, more likely, to be printed on demand at point of sale, in either case for a fee equal to the publishers; normal costs and profit and the authors' contractual royalty, thus for the first time in human history creating the theoretical possibility that every book ever printed in whatever language will be available to everyone on earth with access to the Internet.

The companies that survive the change in consumer habits will be the ones most ready, willing and able to adapt their products into digital formats. In many cases this will entail breaking their products into smaller, more easily consumable bits, which can then be enjoyed by consumers wherever and whenever they see fit.

Publishers could soon be microchunking their texts, separating them into stand-alone, custom-made pieces that can be purchased and consumed individually. If they did, they would have a chance at winning back Generation Upload, who have come to expect a certain if not basic level of interaction with what they consume.

While many of the economic realities of microchunking text and literary content are still being worked out – publishers and agents wrestling with the ideas of what to charge for page-views and partial-downloads of books, and what to pay authors from the revenues received from such sales – consumers are quickly warming to their new abilities. The fact that the creation and consumption of individualized content will be one of the hallmarks of our digital age is not in doubt.

In 1965, when pop songs ruled the airwaves, The Who's second single was called 'Anyway, Anyhow, Anywhere.' This title sounds very much like the anthem for the on-demand everything mindset. If you have an Internet connection and a laptop or wifi handheld device, then you can download and experience just about anything: movies, music, television shows, and soon – hopefully – books. And you can then experience that content anyhow, anywhere you like.

In late 2006, The Who released their first record in over twenty years. For a band whose first releases were on seven-inch vinyl, and whose last album could be bought on a cassette, their new record was sold on compact discs but could also be downloaded from iTunes. Fans not even born when the previous Who record came out (let alone the first one), were able to purchase the songs online and download them straight onto their iPods. The group who once sang 'I hope I die before I get old' stuck around long enough to be part of the future.

7

ebooks and the revolution that didn't happen

IN THE PAST hundred years many predictions have been made about the future and fate of the book that ultimately did not come true – for instance, that paperbacks would be the end of publishing, or that CD-ROMs would revolutionize the business. But the prediction that seemed to have the most weight and hope behind it was the one about electronic books and the impact they would have.

Of course sometimes what's most interesting in terms of technological advancements isn't marveling at what happened, and then forever after admiring the genius and grace of the items and gadgets that have changed our collective lives. Sometimes the real fun comes from looking at predictions of the future and seeing what *didn't* happen. Can any contemporary viewer watch *2001: A Space Odyssey* and not twitter just a little bit at the kooky 'modern' fashions we were all supposed to be wearing at the turn of the century?

If hindsight is 20/20, then our eyesight for looking forward is the exact opposite; instead of clarity, we are faced with an unending horizon that leaves us groping in the dark. We try to make sense of the disparate impressions and diverging ideas, guessing at what we think is going to happen, but we might as well close our eyes and throw darts at a dartboard. We might as well guess. And, after so many wrong predictions about what's going to happen in the next five or ten or twenty years, we may be tempted to ask, as J. G. Ballard already did, 'Does the future still have a future?'

In every book or film or piece of art from the last century that has depicted the future – from Jules Verne to George Lucas – we usually fault it twice: first for the things that didn't come true, and then for failing to see the myriad of changes that did take place.

Back in 1999 electronic books, or eBooks as they became known, were not only going to change everything but they were also going to *replace* everything. Because of this, heady predictions were made during the Internet goldrush of the late 1990s; printed books and the ink-on-paper experience were rapidly on their way out, and digital delivery and consumption would soon be commonplace. The publishing industry was soon going to be rocked to its very foundation, and only a small handful of its original players would be left after the sea change. Traditional publishers were outdated relics about to be sunk.

It didn't happen. After all the claims and predictions, all the industry forecasts and helium-filled hype, eBooks did not take off in any real way. Now, nearly a decade later, print publishing still exists in nearly the same shape and form as it did in the late 1990s. Electronic reading and eBooks are still a minute part of the overall publishing business, and are still trying to gain a foothold in the consciousness of consumers. Since their introduction at the end of the 1990s, consumer response (not to mention publisher participation) has been cautious at best. eBooks have made some inroads in the marketplace, but nowhere near what had been expected.

'Ebooks haven't exactly set the publishing world on its ear,' wrote Walt Crawford in 2006 on the website *EContent*. He cited both the unrealistic expectations held out for eBooks, as well as the reality of what actually happened:

Early projections had print books becoming obsolescent by 2001, or losing half their market to ebooks by then. By 2000, pundits disclaimed any notion that print books would really go away, but ebooks were still a sure thing. In 2001, Accenture projected 28 million dedicated ebook readers in the U.S. by 2005 and $2.3 billion in 2005 text sales for those devices. Forrester projected a more modest $251 million in ebook sales for dedicated readers in 2005 (plus $423 million in ebook sales for other devices) – and another $3.23 billion in digital textbook sales. RCA, making ebook appliances at the time, dismissed Forrester's projections as 'ridiculously low.'

So what happened? And does their rejection by the public and industry constitute a kind of referendum on digital reading in general, proving that print is far from dead? First, let's examine why eBooks have so far failed to take off.

From the very beginning eBooks faced a steep battle in terms of generating and sustaining consumer interest. They were, after all, a whole new way of doing something that people had been doing fine for five centuries. Everyone learns how to read with physical books, and books have been a constant in society ever since. eBooks in the late 1990s were then the answer to a question no one was yet asking.

As early as 2001, David Kirkpatrick wrote in *The New York Times* that 'the tepid demand [for eBooks] comes as no surprise to some bibliophiles, since printed books still work just fine.' There would be many more eulogies for eBooks over the next half-decade, with many critics and industry insiders wondering why there was a call to fix a product no one was yet considering broke.

'About a decade ago, some publishers were predicting that books would soon be a thing of the past,' wrote Charles McGrath in *The New York Times* in 2006, 'and that we would all be reading downloadable texts on portable hand-held screens. Wishful thinking, it turns out.'

What doomed eBooks were the early calls for change by various sectors that did not seem to keep the needs of readers in ind. Leading the charge were a number of companies developing all kinds of amazing technology which, however impressive, was not anything that readers were asking for, let lone clamoring for. During the feverish days of the dot-com oom, the 'if you build it, they will come' mentality permeted everything, including the world of books.

'Sometimes technologists forget just how vast the chasm is etween them and *real* people,' wrote Pip Coburn in his 2006 ook *The Change Function: Why Some Technologies Take Off nd Others Crash and Burn*. 'Many real people resent technolgy. So it won't be easy for technologists to survive this crisis tact – this realization that it is real people and not technolo-ists who determine the fate of technologies.'

People rarely embrace massive change unless they have a lear need or desire to do so. But even keeping this in mind, here were a number of factors that hindered widespread Book adoption, including selection, pricing, format confu-on and digital rights management (also known as DRM). nd, of course, there was even confusion about the name. Vhat exactly *is* an eBook? Is it a format or a device? Is it a phys-al thing, like the Sony eReader? A dedicated device that is sed for just reading books? Or is it the exact opposite, a file ke a Word document or Excel spreadsheet, which is down-aded to a portable device? No one really knew, and this idn't help.

Despite the fact that most of the major New York publishers vere participating in eBooks at the time they made their initial plash, the selection of electronic titles available to consumers

over the years has been woefully small. Most bookstores carr
100,000 titles, while Amazon features well over a million. Bu
even the most well-stocked eBook stores in the late 1990
were lucky to feature around 10,000 titles. Many reader
experimented with electronic reading, but rarely came bac
for more because there just weren't enough titles to choos
from.

Pricing, if a potential reader found something they wante
to read, was also a discouraging factor for many consumer.
The general public felt that electronic books, which were ofte
priced the same as the print edition, should have been muc
less expensive. How much less? There were some calls to cu
the price in half (if not more).

eBooks don't have to be manufactured – after all, every dig
tal copy is exactly the same as the next – and they don't nee
to be stored in a warehouse or be trucked to distribution cen
ters or to stores, so why weren't they sold at a fraction of th
costs of regular books? What about all that money publisher
were saving in paper? Publishers, wary of the threat tha
eBooks posed, didn't want to price eBooks so low that the
would cannibalize the sale of their regular print books. Mean
while, consumers who wanted to pay just a couple of dollar
for the electronic file of a book, faced with paying $27.95 fo
something that 'doesn't really exist,' opted to either pay tha
amount for the printed version or else didn't pay it at all.

Keep in mind that all of this was taking place between 199
and 2001, and the iPod didn't start to gain steam until 2002
Until iTunes established standard pricing of $0.99 per song
and $9.99 for entire albums, consumers were not only facing
numerous business models and various levels of pricing fo
music, but they were also offering stiff resistance. Napster ha
taught a generation that content should be free, while hun
dreds of websites that offered free content, including thos
from for-profit corporations like *The New York Times*, did littl

to convince users otherwise. The publishing industry stumbled into all of this and was instantly met by a cacophony of various voices, each loudly pleading its case: consumers felt that the price for electronic books should be incredibly low, technology companies wanted publishers to drop their prices so that the market would gain traction, and agents and authors – worried about a completely digital future – wanted even higher royalty rates than for print publication.

Adding to the woes of the early market were a number of competing devices and formats vying for the consumers' attention (and dollars). In many ways the early days of eBooks resembled the time when video cassette recorders were first introduced. Instead of choosing between Beta and VHS, readers were asked to choose between Rocket eBook, Softbook, Microsoft Reader, Palm Digital Reader, MobiPocket, and Glassbook. Some of these were dedicated reading devices, some were files meant to be used on personal device assistants such as a Palm Pilot, others were to be read on desktop or laptop computers, and one could be read on almost anything (including cell phones). All of this – for the consumer already unsure of what eBooks were – equaled frustration instead of choice.

The players ranged from typical Silicon Valley garage start-ups to some of the biggest names in the computer industry (including Microsoft, who lurched into the picture with its Reader program in 1999). All of the hype began to attract outside investors, and soon venture capitalists were pumping money into young companies that – while they had some amazing technology – did not yet have a business plan, or were not meeting a need that was going unanswered in terms of everyday readers. Very quickly this became big business, if not a big market.

With each technology company trumpeting its own digital solution, the consumers were left scratching their heads and wondering which was best. And when people are confused, they don't pull out their wallets.

Even when consumers found a book they liked, at a price they didn't mind paying, and then chose the format that made sense to them, they then had to figure out how to download the file to their machine or device. This was often an arduous experience that even a computer scientist would have found confusing. Since electronic books were really just digital files that were transferred to and stored on your computer, which were then synced up and downloaded onto a device, where did the files actually live? How did they get from the website to your device? What could you do when you paid for the title, tried to download it, but then could never find it? Or how about when it was stranded on your desktop with some strange file extension that your computer didn't know what to do with?

Once consumers were able figure all of this out, and managed to get the eBook files from a website onto the device on which they were going to ultimately read it, they were instantly met with another challenge: digital rights management. Also known as DRM, digital rights management is a convoluted set of electronic rules put in place to protect the copyright owners of the material. What it means in practice is that users who have bought the electronic book were not allowed to copy, print, or share the text (not even to another one of their own devices, let alone with someone who might be interested in the material).

Many users of eBooks found the various levels of DRM too constrictive. After all, if you buy a print book with no restrictions as to where you can take it and read it, why should eBooks be any different? Many consumers who purchased eBooks wanted to be able to read the file on a variety of machines or devices in different locations, or else wanted to send it to their friends the same way paperback books get assed on from person to person. But the technology companies – at the behest of the publishers who were afraid of the Napsterization of publishing, fearing that pirated books would be the downfall of their industry – insisted on draconian DRM measures that left early adopters of eBooks feeling shorthanged. In the end, there more things their eBook couldn't do than things they could do.

When eBooks were first introduced in 1999, many people outside the industry were making grand predictions that electronic reading would not only be the wave of the future, but would one day be our collective present; soon, eBooks would be all that everybody knew. Of course, the makers of books breathed a huge sigh of relief when eBooks began to fizzle instead of setting the world on fire. That relief then turned into *schadenfruede*, with many in publishing secretly happy to have (for the time being) dodged the digital bullet. The prevailing emotion seemed to be, 'The Internet is going to have a huge effect in lots of *other* industries, but not ours.'

And because of this – because the eBook revolution *didn't* happen – many contend that the public has already rejected the idea of digital reading. After all, if people wanted to ditch their books for an electronic version, why didn't eBooks take off the way some people had predicted?

This reasoning is wrong because a distinction needs to be made from what became known as 'eBooks' and the act of digital reading itself. True, eBook sales are still minuscule compared with traditional book sales, and yet many of us spend all day reading electronically, including magazine and news material, email and web surfing. More people are consuming information via computer screens than ever before.

Another specific charge often levied against eBooks is that they're not enough like regular books. Most of the early eBook formats and devices tried to faithfully mimic the ink-on-paper experience, and they failed not because they didn't look like real books, but because they looked *too much* like traditional books.

In the 1979 film *Apocalypse Now*, when Martin Sheen's character is watching his fellow soldiers enjoying a Texas-style barbeque in the middle of a jungle in Vietnam, he thinks to himself, 'The more they tried to make it just like home, the more they made everybody miss it.' The same might apply to all of the eBook devices and formats that try to simply mimic the appearance and functionality of a printed book; the more they aim to resemble print, the more people will compare it to a book. And when eBooks contain no searches or hyperlinks, as Blake Wilson points out on *Slate*, 'ironically, it's significantly easier to find information in a paper book than in its digital equivalent.'

It's very tempting to try to keep vestiges of the print book experience in an electronic world, but doing so only does two things inadequately (trying to be both a computer *and* a book) rather than doing one thing well (providing a new way to read). After all, when CDs were first introduced they weren't double-sided in order to mimic vinyl records. Instead, compact discs offered musicians the chance to record eighty minutes of uninterrupted music (rather than forty minutes in two twenty-minute chunks), and many artists jumped at the chance and created new works of art that wouldn't have been possible in the vinyl era.

Novelists of the future will be afforded a similar freedom with a new generation of writers embracing these format-less opportunities in the same way that the demise of the Victorian

hree-volume novel gave way to sprawling modern master-
pieces like *Gravity's Rainbow*. For example, Jonathan Safran
Foer's second novel, the bestselling *Extremely Loud & Incredibly
Close*, features a large number of drawings, illustrations and
ypographic touches. Foer is pushing the boundaries of what
we consider a 'novel,' and in doing so he's paving the way for
uture writers to perhaps use music and animation in order to
enhance their stories.

When Wiliam Faulkner finished *As I Lay Dying* in 1929, he
wanted each of the characters to be represented by a different
color ink. But the publisher balked, declaring it too expensive.
Today, what Faulkner wanted could be easily accomplished in
a digital setting.

The music industry, faced with the rise of digital music and file
sharing craze of the mid-90s, grappled with similar problems.
Obviously some control needed to be given to the consumer,
but how much?

'The idea [with iTunes] was to strike the happy but as yet elu-
sive medium where [music] labels would feel their intellectual
property was protected, and consumers would be able to
make use of the music without feeling as if they purchased a
disabled product,' wrote Steven Levy in his 2006 book *The
Perfect Thing:*. 'At that point no one was sure that this zone
existed.'

And yet Apple not only discovered this zone, but ultimately
generated billions of dollars from it. Since the advent of the
Pod, the sales of physical albums have plunged, while the
sale of digital downloads has consistently and significantly,

increased. In 2006, 582 million tracks were sold, and 3̄
million digital albums were sold, a 65% increase over the pre
vious year.

In February of 2007, in a stunning move, Apple co-founde
and CEO Steve Jobs – writing on the Apple website – acknowl
edged the various problems that music swaddled in DRM
poses, and asked the technology companies and record label
to embrace a DRM-less future.

'Imagine a world where every online store sells DRM-fre
music encoded in open licensable formats,' wrote Jobs. 'I
such a world, any player can play music purchased from an
store, and any store can sell music which is playable on al
players. This is clearly the best alternative for consumers, and
Apple would embrace it in a heartbeat.'

Even though Jobs took some flak from hard core technolo
gists who felt that iTunes was creating a Windows-like
monopoly in the digital music space, this was an astonishing
move. While a lessening in DRM restrictions (not to mention
getting rid of DRM itself) had been called for by a number o
people and organizations, this was the first time it had been
said by someone as powerful as Jobs. And it sent an earth
quake through a number of industries because, if the majo
record labels agreed to release music without any form o
DRM, then it meant that TV and film companies (and maybe
even publishers) might one day do the same. And of course
once consumers are given the ability to share their legally
downloaded entertainment, not to mention consume them
on any device or computer, or access them from anywhere in
the world, it would be a huge boon to digital delivery and the
large-scale adoption of completely format-less entertain
ment.

It did not take long for a record label to respond to Jobs's
challenge. In April, EMI – the world's third largest record com
pany – announced that it would begin selling music from its

artists as digital downloads without any kind of DRM or copy restrictions. The songs have a higher sound quality than a typical iTunes download, but cost $1.29 instead of the usual $0.99. The company made the decision after hearing numerous complaints from its consumers that they preferred having format-less music that could be listened to on any computer or any device, using a multitude of programs. Eric Nicoli, EMI's chief executive, was quoted in *The New York Times* as saying 'It was clear what we had to do because we hold the consumer at the center of our focus.'

This was a dramatic move, and showed that EMI wants its music to reach as many ears as possible, while the Recording Industry Association of America has continued to go on the warpath and sue music fans for the illegal downloading of music; the RIAA would rather create defendants instead of an audience.

The decisions about DRM that are now being made in the music world will mean a lot for trade publishers and eBooks in the years to come. After all, one of the big problems which has restricted eBook adoption is the restrictions of DRM. True, most publishers are only reacting to authors and agents who are very leery of digital delivery (and the devilry they fear it will bring: copyright theft, loss of revenue, mass piracy). But compared to the alternative – no one wanting to read their books – it's apparent that the time has come to experiment and put the power in the hands of the consumers.

What was also significant about the EMI decision is that the price-per-song was higher, proving that people will (hopefully) be willing to pay for the convenience that digital delivery provides. A price of $0.99 locks you to your iPod, but $1.29 lets you take it anywhere you want. This could one day be the same for books. So instead of electronic books being priced ridiculously low (as some people have called for, wanting eBook prices to be somewhere in the $1–$2 range), consumers will instead pay comparable if not premium prices for digital downloads of books. Why? Because – if the files are not straitjacketed with DRM – then the users can read the files on any device or on any computer, at any time or in any place that they want. This could prove liberating, and would finally be one in the win column in the 'books vs. eBooks debate, since a digital file is a virtual item that can live in many places at once, while a printed book is a physical thing that has to be dragged by hand from place to place.

It remains to be seen how many of the other major record labels – if any – step forward and make a step similar to EMI's, but at least it's a start. And in terms of books, no one knows whether an iPod-like device for reading will emerge to catalyze rapid change. Certainly there are contenders for just such a position. Every eBook device that has been developed in the past few years, is compared to the iPod, with industry insiders watching and wondering if one of them will cause electronic reading to explode.

While the iPod – and the change in habits it has spurred – is a tremendously good thing, it's also proving a distraction for the publishing industry. Executives now keep asking themselves, 'When will the iPod for books arrive?' Indeed, in a *Business Week* article about the debut of Sony's eBook device, 'Apple' and 'iPod' are mentioned ten times. This wasn't the case in the first round of media about eBooks circa 2000, since the iPod had not yet been invented. What was also missing from those articles was the pressure for any of the proposed business models to achieve what Apple has achieved in less than five years. So eBooks, as they currently stand, can't help but look like an also-ran next to MP3s. Expecting them to perform in the same sales league is like expecting lightning to strike in the same place twice.

Instead, what has a better chance at revolutionizing electronic reading may in fact *not* be an iPod-like dedicated device for reading. Real success will probably be achieved through seamless integration with a piece of existing hardware, something people are already carrying around with them, such as a PDA or wireless tablet. This kind of integration has already become commonplace (for instance, the combination of cell phone and digital camera). The Apple iPhone – with its ability to play music, videos, TV shows, display photos, make phone calls and surf the Web – points the way to the super-device of the future.

Therefore, even though eBooks failed to catch on a few years ago, we need to realize that it's not only information that travels fast in a digital age; habits change fast, too. And it's wrong to say that, since people didn't warm to electronic books in the late 1990s people will never warm to them. Remember that iPods weren't introduced until October 2001, and it then took less than four years for them to be, well, *everywhere*. And of course iPods were just another MP3 player, many of which had been on the market for years. And if habits, tastes, and behavior towards music can change that quickly, then habits, tastes and behavior towards reading can change that fast, too.

'If nothing else, futurologists do have the habit of announcing both deaths and births prematurely,' wrote Paul Duguid in the 1996 essay collection *The Future of the Book*. 'Talking machines, domestic robots, automated language translators, and a host of other "new technologies" have, for forty years or more, been perennial examples of "vaporware," always coming yet never coming "within the next decade".'

So while the futurologists and the pundits were wrong in predicting the dominance of electronic reading a decade ago, what's happened since has been the explosive growth of the always-on Internet. The digital habits of just about every sector of society have shown that things *did* change. Just because the digital tide stopped at the feet of publishing doesn't mean the flood's not coming; it only means that the water's getting close.

Consumers, of course, were only one segment that had to get interested in electronic reading. What about publishers? For eBooks to have any chance of succeeding, major publishers would have had to make huge commitments to electronic reading. They would have had to invest millions in electronic infrastructure and digital conversion. And since eBooks were really just a prediction back in 1999 – and not a proven business model – many publishers were cautious, and committed limited resources to their eBook programs.

'Publishers and online bookstores say only the very few best-selling electronic editions have sold more than a thousand copies,' wrote David Kirkpatrick in *The New York Times* in 2001, 'and most sell far fewer. Only a handful have generated enough revenue to cover the few hundred dollars it costs to convert their texts to digital formats.'

This quickly turned into an eBook Catch-22: sales are small, so publishers don't convert and make available their entire catalog. Yet sales are small because there's not a big enough selection from the publishers of books to choose from.

In addition to this, the very idea of eBooks tends to make publishers uneasy. While almost every major trade publisher took part in some way in eBook experiments in the late 1990s, and some – including HarperCollins and Random House – started expensive eBook initiatives (including online stores and digital-only imprints), the efforts of most publishers were reserved to say the least. Why? Because if eBooks were successful, they would then drive down the need for their other products and services.

This thinking has occurred before in numerous industries. For example, why won't the cigarette manufacturers produce the more healthy smokeless cigarette that they've long been developing? Because to do so would point out how unhealthy their regular products are. Why doesn't McDonald's have at least one low-fat burger on their menu? Or make a kind of French fry that won't clog up your arteries with grease? Because to do so would mean admitting that everything else they're selling is bad for you. The bottom line is that tobacco companies and fast food franchises have each already made billions of dollars off consumers who purchase things that are bad for them, and the production of an alternative would be a tacit admission that these companies had known all along that what they were doing was wrong.

So while publishers have indeed participated in eBooks since the late 1990s, not many of them had their hearts in it. This is because they have so much staked on print-on-paper books that if eBooks succeed – on almost any level – then their entire way of doing business will be jeopardized.

Not to mention the fact that, at its tweedy heart, publishing considers itself an old world industry indebted more to the gilt edges of Gutenberg than the smooth finish of Apple. Culture plays as big a part in these decisions as business does, and book culture is – by its very nature – retro.

saying goodbye to the book

8

writers in a digital future

George Gissing's novel about the down-and-out literary life of London, *New Grub Street*, paints a brilliant portrait of authors struggling to write meaningful material in a marketplace that relies on gaudy bestsellers revolving around salacious subject matters such as murder and sex. At the center of Gissing's numerous characters grappling with the question of art versus commerce, and trying to make a living doing what they love, is Edwin Reardon. Reardon is a poor-selling novelist doing everything to matter in a literary world that seems to be all about business.

'Literature nowadays is a trade,' says Jasper Milvain, an ambitious young friend of Reardon's who criticizes him for not doing more to promote his books. 'Putting aside men of genius, who may succeed by mere cosmic force, your successful man of letters is your skilful tradesman. He thinks first and foremost of the markets; when one kind of goods begins to go off slackly, he is ready with something new and appetising.'

Gissing's novel was published in 1891. Over a hundred years later the book and its concerns are still poignant and fresh. Writers today have the same struggles, plus a myriad of new entertainment options to battle against for consumers' time and interest. One hundred years on from *New Grub Street*, the general public has less and less time to read for leisure.

So while literary culture is undoubtedly threatened by the rise in computer culture and the diversions it represents, our digital age will also offer authors many opportunities in terms of exposure and marketing – potentially reaching new audiences and interacting with readers – not to mention fostering new styles of writing and ways of reading. All of which adds up to the biggest change for writers since the days of Gissing.

Technology has always played a role in the creation of reading material. In the fifteenth century, Gutenberg's invention allowed, for the first time, information to travel in different directions at once. Before the printing press, the various versions of books or recorded information (the illuminated manuscripts of the Middle Ages, the codex of the Roman Empire, the papyrus sheets of the Egyptians and, where it all began, cuneiform pressed into clay tablets in 5000 BC) consisted of handmade, one-of-a-kind objects. One person with information would pass that information to another person, using the book as a baton. What Gutenberg's printing press, along with movable type, finally allowed was for information to be transferred from one person to many different people at once. As Victor Hugo wrote in *The Hunchback of Notre Dame*, 'In the fifteenth century, everything changes.'

It would be another couple of centuries before, with the appearance of the typewriter in the late 19th century, technology finally had a hand in the way that books were written. Before then, all of the major technological innovations were on the publishing side, while the innovations in terms of writing had mainly to do with new ways of telling stories (stream of consciousness etc.). But fifty or sixty years ago, when most writers began to use typewriters, there was finally a pronounced shift in the way writers thought about and composed their work.

'The typewriter fuses composition and publication, causing an entirely new attitude to the written and printed word,' wrote Marshall McLuhan in *Understanding Media*. 'Composing on the typewriter has altered the forms of the language and of literature.'

Twenty-five years ago the invention of the personal computer, and the development of sophisticated word processing programs that went along with it, altered those forms even more. Writers were suddenly freed from much of the drudgery

of the writing process in the same way that household appliances like dishwashers and vacuums saved the housewife of
the 1950s from domestic chores. Writers could now repag-
inate or change a character's name throughout the entire text
with just a few keystrokes. Countless hours were saved, and
the processes of composition and revision were incredibly
simplified by writing on computers.

'I wrote my first novel on a big clunker of a machine that
wheezed slightly when it stored information and had a mere
256 kilobits of memory,' wrote Anna Quindlen in her 1998
book *How Reading Changed My Life*:

> It just managed to hold the book, the word-processing
> program, and a few other odds and ends. My third novel
> was composed on a machine that fits into my handbag
> and weighs slightly more than a premature baby. The
> program corrects my punctuation and capitalization as I
> type; when I try to type a stand-alone lowercase *I*, it
> inflates it into a capital letter, correcting me perempto-
> rily, certain I've made a mistake. I could keep a dozen
> copies of my book on its hard disk and it wouldn't even
> breathe hard.

So if writers can spend months and years *composing* on a
computer – not only reading words on a computer screen, but
writing them as well – then it's not unrealistic to think that
most readers will one day also consume those books on some
sort of electronic screen. Some may see this as a leap of logic.
As a friend of mine pointed out, milk is produced by squeezing
a cow's teat but most people wouldn't want to consume it
that way. Others cite our willingness to read large amounts of
material online via websites and blogs, online-only magazines
such as *Slate* and *Salon* (and the online editions of traditional
print newspapers such as *The New York Times*) to show that

words produced on computers will soon be consumed on a computer as well.

For authors, this will not be a stretch at all. In fact, they're already used to it. Many authors these days write their books on a computer, and never see their work as a physical manuscript until it's time to send it to their agent or publisher. And even then they might not print out a paper copy, emailing instead electronic files. In fact, some authors *never* see their books in paper format until right before the book is printed.

Books used to be a kind of facsimile of the writer's creation; the writer typed and the publishers turned that type into recreations called books. Today's writers are composing on computers, and then those bits of data are being transformed into physical books. It's the digital versus analog debate turned inside out. Because it used to be that something analog (like a book) was turned into something digital (like an eBook), and people saw this as false. In this scenario, eBooks were seen as being like a clockwork orange: something mechanical pretending to be organic. But now we have the opposite; orange clockworks: something organic created from something mechanical.

However, while such a leap in computer technology has made life easier for writers, it also presents a very large downside: one computer crash, and – unless you back up your work relentlessly – your book and whatever else it is you're working on will disappear in the blink of an eye. Weeks, months, or even years of work can disappear in a matter of seconds, and digital obliteration lies just an incorrect mouseclick away.

This reminds me of an episode of the television show *Mad About You* from years ago. Paul Reiser was fooling around on the roof of his Manhattan apartment building when he knocked out the power. After he goes back to his apartment, he's greeted by a gathering of angry tenants, one of whom is a writer. Apparently the writer had been working on a book at the time of the outage, only to lose all of his work during the ensuing blackout. He hands Reiser a floppy disc. 'What's this?' Reiser asks. 'It used to be my novel,' the writer responds, 'now it's a coaster.'

So while the young artists of *La Bohème* destroyed their art by feeding manuscripts to the eager fireplace of their garret in order to keep warm, writers today can erase their life's work by tripping over a power cord and knocking the plug from an outlet. It's not nearly as poetic as Puccini, but the results are unfortunately the same.

With nothing more than a laptop and a program like Garageband, a young musician alone in his bedroom can lay down unlimited tracks and access a whole library of sounds, effects and synthesized instruments. Forty years ago the Beatles made *Sgt. Pepper's Lonely Hearts Club Band* on a four-track tape recorder. Musicians today can easily layer dozens upon dozens of tracks (and by using digital technology, there's no tape hiss or loss of quality when copies are made).

Similarly, inexpensive computer animation and digital camcorders has given filmmakers access to locations and special effects that previously only the biggest of budgets could afford. Cameras are inexpensive, and digital filmmaking is becoming more and more accepted among cineastes. The old distinctions of 35 mm versus 16 mm (let alone 70 mm, which used to be the holy grail of filmmaking) have all been erased with the introduction of digital filmmaking. Additionally, a young group of filmmakers is beginning to eschew film school, opting instead to make movies on their computers. Why? Because they can.

In 2003, one such filmmaker, Jonathan Caouette, made a documentary about his life named *Tarnation*, putting together the movie on his iMac at a cost of only a couple hundred dollars. The film went on to become a critical and commercial success, garnering several awards and competing in major film festivals.

Computers will also play a part in the new kinds of works writers will produce. In the same way that technology has changed the way musicians and filmmakers record songs and make movies, so too will it change the types and style of narratives that writers of the future will fashion.

Some writers and computer enthusiasts have already been experimenting with these concepts, dating back to the invention of 'hypertext' in the mid-1960s. Used to describe literary works which are computer-based and which contain and link not only to text but also to music, photographs, and any other kind of multimedia experience, hypertext opened a whole new area of literary experience. Since its introduction, hypertext has been consumed by a mostly cult following, created by experimental writers existing far outside mainstream publishing.

One of the most successful long-form literary hypertexts is Michael Joyce's *afternoon: a story*. Written in 1987 using a program called Storyspace, *afternoon* gives readers the ability to make decisions, at numerous points in the story, that impact the action. This allows the narrative to unfold at the fingertips of the reader, changing with each encounter like going through a maze and choosing different points at which to turn left and right.

'Reading *afternoon* several times is like exploring a vast house or castle,' wrote Jay David Bolter in his 2001 book *Writing Space: Computers, Hypertext and the Remediation of Print*. 'Although the reader may proceed often down the same corridors and through familiar rooms, she may also come upon a new hallway not previously explored or find a previously locked door suddenly giving way to the touch. Gradually, she pushes back the margins of this electronic space – as in a computer game in which the descent down a stairway reveals a whole new level of the dungeon.'

These ideas were even partially explored in a series of young adult novels popular in the early 1980s. Entitled *Choose Your Own Adventure*, these short books let readers, throughout the course of the story, make decisions about which way the plot would twist and turn. At multiple points in the book, readers made decisions based on choices in the story, such as did they want to enter a house or else bypass the house and go to the one next door?

Each book was written in the second person point-of-view (an underused but effective literary device), which made the reader feel as if everything was happening directly to him or her (or rather, to *You*). Depending on the choices the reader would make, they would go to a different part of the book and continue reading the story that they had created. Each book had dozens of possibilities, and a variety of endings. Readers, myself included, found the experience intoxicating. The *Choose Your Own Adventure* series went on to sell over 250 million copies, becoming one of the most successful children's book series of all time.

However, fun as they were, reading a *Choose Your Own Adventure* book was exasperating; every time one came to a point in the story where a decision had to be made, one had to follow instructions on the bottom of the page (for instance, 'If you want to enter the house, turn to page 26'). All of this flipping around felt like trying to find a word in the dictionary, taking the wind out of any narrative drive that had been building up.

If these had been electronic stories, digital words on a touch-sensitive screen, using hyperlinks in the text ('If you vant to enter the house, touch here') readers would have jumped to any other part of the book instantly, creating a seamless and instant experience. The story would unfold at the touch of our fingertips, as if we truly were the creator.

Alongside the *Choose Your Own Adventure* books, even more primitive text-based video games were being offered by a software company named Infocom. In particular, Infocom's Zork series was immensely popular with computer enthusiasts of the early 1980s. Completely text-based, these games would serve up short informational nuggets ('A path leads northeast through this small grove of orange trees. A dark cave lies to the west. A sign is posted near the grove.'), with the user then having to take action in order to move the story along. These appealed to the Dungeon and Dragons crowd, and had a hard core group of fans for a number of years. However, the company fizzled when the graphics of home computers allowed for more realistic, first-person adventure games. Looking back, both the Infocom games and the *Choose Your Own Adventure* books were early versions of the role-playing environments that Generation Download now immerse themselves in for hours at a time (such as *World of Warcraft* and *Second Life*). And even though these early examples may today seem pretty pedestrian, at the time they offered a tantalizing glimpse into the future of text-based games and interactivity.

Also, while both the *Choose Your Own Adventure* series and the Infocom games were crude approximations of what the Internet would become a decade later, the level of interaction they provided paved the way for the kinds of literary and narrative experiences future readers and computer users could achieve. For instance, what if users could shuffle the chapters of books and make their own literary remixes? Or else, authors could provide alternative edits or versions of their books, one version featuring an emphasis on one character while a change in the settings would make yet another character the protagonist. Mystery writers could write alternate endings to their novels in which a different character was murdered each time you read, creating a *Rashomon* effect that would give new color to all of the chapters that preceded the altered endings.

While hypertext and its nearly unlimited potential for weaving in various strands of multimedia and user interaction is seen by a small number of writers today as a liberation, many others perceive it as a dire threat. For legions of writers who have been inspired by and schooled in the classics of the 19th and 18th centuries, embracing the techniques of hypertext will be challenging and difficult.

'Writing students are notoriously conservative creatures,' wrote Robert Coover in *The New York Times* in 1992. 'They write stubbornly and hopefully within the tradition of what they have read. Getting them to try out alternative or innovative forms is harder than talking them into chastity as a life style. But confronted with hyperspace, they have no choice: all the comforting structures have been erased. It's improvise or go home.'

Of course, some of these ideas are not only decades old, but have existed for hundreds of years. Laurence Sterne's groundbreaking and hallucinogenic 18th century novel *The Life and Opinions of Tristram Shandy, Gentleman* – with its short, choppy chapters and rambling framework (even though the book is ostensibly about the life of Tristram Shandy, the title character isn't born until the third volume) – practically begs for the reader to skip around the text and read the material out of order.

'In Sterne's novel we find a very particular use of layout and printing devices,' wrote Luca Toschi in the 1996 essay collection *The Future of the Book*. 'We can find a black page, a marbled page, squiggly lines used to express ideas and feelings.

Sterne seems to have wanted to make people think about the expressive capacity of words and of the ways of representing these through this eccentric use of printing.'

Two hundred years later, in the 1960s, Norman Mailer also played with the concept of offering the reader different ways to experience the same book. He did this most explicitly with 1961's *Advertisements for Myself*, a collection of assorted pieces of journalism as well as short stories and introductory passages. Mailer offers up two tables of contents, giving the reader dual ways to explore the book. 'The First [table of contents] lists each piece in sequence, and anyone wishing to read my book from beginning to end may be pleased to hear that the order is roughly chronological.' However, 'A Second Table of Contents is offered to satisfy the specialist. Here essays, journalism, and miscellany are posted in their formal category.' Mailer then offers even a third way to read the book, listing what he considers to be the best pieces of the book, 'for those who care to skim nothing but the cream.'

In another Mailer book, *Armies of the Night* – his Pulitzer Prize and National Book Award winning account of a 1968 anti-war demonstration in Washington DC – the story is divided into two separate sections. Subtitled 'History as a Novel, the Novel as History,' Mailer offers the reader two ways to view the material, one in which he concentrates on the historical facts of the demonstration, and yet another in which Mailer emphasizes his more novelistic approach.

While both of these books offer intriguing ideas, the real world application (in terms of *Advertisements*) – like with all of the flipping back and forth of the *Choose Your Own Adventure* books – is exhausting. But the opportunities inherent in Mailer's construction and concept are perfectly suited to a digital experience. For instance, imagine an electronic version of Mailer's dizzying maze of prose; being able to choose at the outset the way you wished to experience the content ('Touch

here to read history as a novel, touch here to read a novel a
history'), the same content mixed up two different ways fo
completely different experiences.

Martin Amis is another mainstream writer who has pushed
the boundaries of the print experience. In his 1991 nove
Time's Arrow, the events occur backwards, with scenes and
even dialogue unspooling like a film played in reverse. It's a
dizzying concept, and one that's ripe for digital experimenta
tion and remixing.

Dozens of more experimental texts exist by writers such as
Borges, Calvino, Cortazar and even Nabokov; novels and
short fiction composed of narrative labyrinths and linguistic
gymnastics that push the reader to interact with the text
through a dizzying array of literary techniques and visual
devices. For these artists – and the numerous postmodern
writers they have influenced, such as David Foster Wallace
Nicholson Baker and Donald Antrim – the passive reader does
n't exist. The words are building blocks for readers themselve
to construct.

Musicians have recently offered similar experiments, giving
users the choice to listen to alternate mixes of songs or differ
ent kinds of stereo configurations (such as the recent rise in
5.1 mixes, which are DVDs created for home theater systems)
Many artists and bands – from Nine Inch Nails to David Byrne
and Brian Eno – have set up websites where they give fans the
ability to remix their songs any way they want.

Filmmakers have gone even further, using the technology of
DVDs to allow users to watch different edits of the same

movie. Users can also toggle back and forth between a number of audio commentaries, alternate camera angles, or even go to the shooting script to see how much of the page made it onto the screen. Christopher Nolan's 2000 film *Memento* is a groundbreaking DVD, in which users must figure out clues to unlock the disc's various secrets. One of the features of the disc gives the user the ability to watch the film – which is constructed in reverse chronological order, beginning with the ending and proceeding in short scenes to the beginning of the story – in reverse order (i.e. from beginning to end). Imagine if the readers of *Time's Arrow* had the same opportunity, re-ordering all of the words in the book so that they could read it in chronological order.

The Beastie Boys combined all of these elements on the Criterion Edition of their collection of videos, allowing users to choose a video and then further choose from different audio mixes in addition to video edits and other bonus features. This allows the user to create their own music videos. Over the span of 18 videos, according to the Criterion website, 'there are hundreds of possible image and sound combinations, including new surround mixes, a cappella versions, instrumentals, and more than 40 remixes.' I myself have owned this DVD for a couple of years, and have spent many hours with it, and have still not played every available combination. Every time I watch it, the videos give me something new, something I hadn't experienced before.

'It is the artist's job to try to dislocate older media into postures that permit attention to the new,' wrote Marshall McLuhan in *Understanding Media*. 'To this end, the artist must ever play and experiment with new means of arranging experience, even though the majority of his audience may prefer to remain fixed in their old perceptual attitudes.'

Right now, the book-buying audience is indeed 'fixed in their old perceptual attitudes,' but this cannot but change in time. The amount of user interaction we've seen at the hands of Generation Upload proves that many are already there. Because of this, and with digital technology – and the demands by future readers for an interactive experience – the boundaries of narrative text will be pushed even further than the examples already cited.

The Internet, and the interconnectedness of its audience, is already proving to be a great place for growing and sustaining an audience online. Some authors today have huge and ravenous Internet constituencies, and in terms of their popularity the lines are so blurred between their websites and their books that it's no longer clear which exists to support the other. And all of this comes at a time when the book business and authors are faced with more competition than ever in terms of consumer attention and competing fields of entertainment.

'The book business has always been a tough business but never as tough as it is here and now,' wrote Pat Walsh in his 2005 book *78 Reasons Why Your Book May Never Be Published and 14 Reasons Why it Just Might*:

> Americans do not buy a lot of books compared to other countries such as Germany and Spain. When they do, they tend to buy the same author over and over – particularly novelists. Recent surveys have shown that the readership for books is down, and that trend does not look likely to abate anytime soon. This does leave a lot of room for writers to break in and it certainly makes earning a living as a writer a near-impossibility.

Here's where the Internet goes from looking like print's killer to its savior, offering tremendous opportunities for writers who are willing to work to harness its power. Author Chuck Palahniuk is a great example of this. His fans started a website about him in the late 1990s, and since then not only has a huge community grown up around it, but Palahniuk has personally endorsed the site. In 2003 the members of the website even made a documentary entitled *Postcards From the Future: The Chuck Palahniuk Documentary* (which of course can be ordered from the site). The webmasters also organize conferences devoted to Palahniuk and his work, and the website includes a writer's workshop (during which Palahniuk offers tips and advice). The site also sells Chuck Palahniuk merchandise, accepts advertising, and now charges members to access all of the site's various features.

In the same way that Jimmy Buffet has created a multimillion dollar business around the success of his 1977 song 'Margaritaville,' so too will future authors create online communities and brands built around their works that have the potential to be even more popular than the works the communities were built to support. The Internet itself is developing into a very effective marketing tool for publishers and writers, allowing them to inexpensively get exposure for books to exactly the people who might want them.

For example, for most writers a book tour has always been an impossibility. It costs a lot to get an author from city to city to sign a couple of books in a few malls. And the benefits – for even the biggest authors – are often negligible. Instead, writers can do a 'virtual tour' from the comfort of their homes, for no money, visiting dozens of literary websites and blogs, connecting with thousands if not tens of thousands of potential readers.

'We think of books through a commercial lens, assuming that most authors want to write a bestseller and get rich,' wrote *Wired* editor Chris Anderson in his influential bestselle *The Long Tail.* 'But the reality is that the vast majority c authors not only won't become bestsellers, but also aren' even trying to write a hugely popular book. Each year, nearl 200,000 books are published in English. Fewer than 20,00 will make it into the average book superstore. Most won't sell. This is what writers are up against, and the Internet can help.

Perhaps the best example of an author at the forefront of nev online marketing and self-promotion techniques to expos themselves to an Internet audience is cutting-edge science-fic tion writer Cory Doctorow, a self-avowed 'techno-agnosti social-democrat libertarian.' Doctorow, who is also active in th debate over the future of copyright, as well as being one of th founders of the popular website Boing Boing, has offered ever one of his books as a free download from his website, allowinç users anywhere in the world to access his work for nothing Instead of driving down the sales of his physical books Doctorow insists that the exposure the free downloads give him spurs sales instead of detracting from them.

'Most people who download the book don't end up buying it, but they wouldn't have bought it in any event, so I haven' lost any sales, I've just won an audience,' wrote Doctorow in *Forbes Magazine* in 2006. 'A tiny minority of downloaders treat the free e-book as a substitute for the printed book – those are the lost sales. But a much larger minority treat the e-book as an enticement to buy the printed book. They're gained sales. As long as gained sales outnumber lost sales, I'm ahead of the game. After all, distributing nearly a million copies of my book has cost me nothing.'

Doctorow has also released his books under a Creative Commons License, which legally allows – and indeed, encourages users – not only to distribute his work but also to use it in new and interesting ways, including translations or multimedia interpretations.

All of this has led to a wildly successful writing career in which Doctorow has sold hundreds of thousand of books and been translated into dozens of languages. In addition, this profile has led to innumerable writing and non-writing opportunities (such as magazine assignments and speaking engagements). Doctorow is the poster-boy for a new generation of online-savvy artists who have turned their presence on the Internet into real-world success.

What about the writer who wants only to, well, write? What about those writers who may use the Internet for research and email, composing their books on a computer, but don't want to be part of the digital revolution? What will happen to them?

Authors who choose not to take part in any sort of online promotion or to curry online exposure, and unwilling to do things like start a blog, post clips on YouTube, have a page on MySpace or otherwise engage an Internet audience in any meaningful way will find themselves at an increasing disadvantage.

To begin with, they will have a harder time getting a book deal in the first place. Publishers will be increasingly unwilling to sign authors who do not already have an Internet audience, or have no desire to do any online promotion or outreach. Many bloggers have received book deals precisely because of their online constituencies, with publishers hoping that their large Internet audiences will purchase a book by these often first-time authors. For a prospective publisher, this is a safer bet than signing up a first-time author who is starting to build

an audience from scratch. Blogs can offer a track recor
showing that the writer can indeed connect with readers an
that people are interested in his work. For the writer witho
any kind of recommendation other than his manuscript, it
going to be a much harder sell.

The authors who do get the book deal, but don't want t
engage in any online promotion, will find themselves increa
ingly at a disadvantage when they're competing with book
written by authors who *are* engaged in numerous Intern
activities, such as blogging, communicating with existing o
potential fans, or even giving away large portions of the
book for download or online reading. With fewer reviews o
books appearing in newspapers and magazines, the Internet
the place where readers are finding out about new books. S
books that have either a subject matter or an author attune
to the workings of the online world have a much bette
chance for exposure and visibility.

Competition among writers and books has always bee
fierce, and while the Internet offers new opportunities fo
authors, it also represents a new arena in the struggle to fin
and keep readers. A hundred years ago, in the time of *Ne*
Grub Street, the main thing driving sales was favorabl
reviews. Gissing wrote:

> Speaking seriously, we know that a really good book will
> more likely than not receive fair treatment from two or
> three reviewers. Yes, but also more likely than not it will be
> swamped in the flood of literature that pours forth week
> after week, and won't have attention fixed long enough
> upon it to establish its repute. The struggle for existence
> among books is nowadays as severe as among men.

Yet the world of book reviewing, as mentioned in Chapter 3,
is in disarray. Fewer reviews are being written, and the ones

hat find a home in a newspaper exist alongside websites, blogs, message boards, chat rooms, forums and wikis. If you subscribe to the notion of the 'attention economy,' and agree that books now compete not only with films and television for cultural relevance and interest, but now must also beat the array of new media that the Internet has spawned for human bandwidth, then why shouldn't authors use every marketing tool at their disposal? Why shouldn't they make trailers for their books and videos of themselves?

People criticize this because it will, they fear, change the landscape of the world of literature and give an unfair advantage to some. Short films and videos would be posted all over the Internet, embedded on blogs and traded on file-sharing websites, and all of a sudden a charismatic but not-very-talented writer would find himself more popular than a wonderful writer who freezes up in front of a camera. But this is nothing new. Writers like Norman Mailer and Truman Capote were the first to regularly appear on television talk shows, becoming as famous (or, in the case of Mailer, as infamous) as rock stars. Later generations of smart, attractive writers such as Jay McInerney and Bret Easton Ellis continued in this tradition, appearing in newspapers and magazines as the cover stars of a new literary generation.

The Web will take this even further. And writers who are unskilled in the ways of the Internet, or just don't want to play any part in the online discussion and want to write their books and be left alone, will be like movie actors at the end of the silent era who were forced to have elocution lessons when talking pictures were suddenly the brand new thing. For some, the advent of sound allowed them to shine in a way that silent films never did. Others, however – those who didn't have good voices or couldn't act in the way that talking pictures demanded – found themselves suddenly without a career. Many modern day writers will find themselves in similar circumstances, unable to deal with the ramifications and changes that a new technology has brought to their art form.

True, there will always be some major writers who do nothing but write, but they will be exceedingly rare and they will appeal to smaller and smaller audiences as time goes on. Literary fiction has always been a low-selling, specialized taste – with critically acclaimed writers selling only thousands of copies – and a rise in digital reading and culture will only further marginalize literary fiction. In addition, fewer young writers will follow in the footsteps of past masters like Carver and Cheever, choosing the quicksilver circuitry of the Internet over the quiet anomie of the suburbs. Indeed, those writers who insist on aping the style and convention of previous literary genres and schools will find themselves only isolated instead of praised. Just because we read and admire Henry James today doesn't mean that contemporary writers should try to write like him.

'Let us imagine a contemporary composer writing a sonata that in its form, its harmonies, its melodies resembles Beethoven's,' wrote Milan Kundera in his 2007 book *The Curtain: An Essay in Seven Parts*. 'Let's even imagine that this sonata is so masterfully made that, if it had actually been by Beethoven, it would count among his greatest works. And yet no matter how magnificent, signed by a contemporary composer it would be laughable. At best its author would be applauded as a virtuoso of pastiche.'

In terms of avoiding publicity, writers like Philip Roth have over the years been reclusive, refusing to do interviews or play the major media game. And Roth has managed – in spite of what would be a handicap for other writers – to have as successful a career as any modern writer. But writers of his caliber are rare, and Roth has never had to rely solely on sales; he is a prestige author, a name publishers will pay handsomely for just to have him on their list and in their catalogs. In the case of someone like Thomas Pynchon, whose hide-and-seek with the public provides a mythic subtext to his works, his non-appearance in public is just as potent – if not more so – than if he gave interviews and toured. Roth and Pynchon, and writers like them, will increasingly be seen as exceptions to the emerging rule.

The midlist writers of today, facing an onslaught of intense competition, not just from the thousands of other midlist authors out there, but from all the other kinds of entertainment – which *are* promoting themselves online – are going to find themselves pressured to take part in the conversations taking place online. Similarly, in terms of their audience, the fans of these writers will expect some sort of access to the authors that they choose to read. The age of the aloof writer, removed from his audience or not even knowing who his audience is, is long gone.

New writers will have to embrace not only new techniques of online promotion and participation, but will also have to embrace the new literary forms which digital reading and delivery make possible. Because of this, the Internet will kill as many careers as it gives birth to; for every blogger who is given a book deal, another novelist who simply wants to tell stories will be unable to get a contract. Cumulatively, all the changes of a digital world will transform the experience of being a writer so much that the profession may not resemble in the future what it is today. This is, unfortunately for some, the new truth of our current literary age, and nothing can bring us back to the era of Grub Street.

9

readers in a digital future

READERS SHOULD have all along been the focus of the debate over the future of the book, and yet they have heretofore been relegated to the sidelines. Instead of thinking of ways to give readers a different or more pleasurable experience, many major publishing and technology companies have instead told readers in a patronizing way what's good for them. And bibliophiles, who seem to care more about books than readers, have long insisted on the sanctity of paper and cloth over any kind of utility of text.

'Part of the great wonder of reading is that it has the ability to make human beings feel more connected to one another,' wrote Anna Quindlen in her 1998 book *How Books Changed My Life*, 'which is a great good, if not from a pedagogical point of view, at least from a psychological one.'

The danger for books and writing in the twenty-first century is that humans are already more 'connected to one another' than they've ever been. A hundred years ago we might have read *War and Peace* in an effort to see what life is like half a world away. With the world now joined in an electronic web, anyone curious about life somewhere else on the planet need only spend half an hour online to gain insight into another place. Graham Greene's *Journey Without Maps* becomes increasingly anachronistic in a world where Google Earth has inventoried nearly every backyard on the planet.

Bookstores, which used to be local businesses (even in the age of the megastores), have given way to websites like Amazon, where readers from all over the country can buy and shop for books whenever they like. In fact, the Internet will change more than the way readers read books. It will also – and already has – changed the way people buy, learn about and discuss books.

There are now numerous social networking sites devoted to books, among them Shelfari, Good Reads and Library Thing. At these sites, users create profiles and construct 'virtual bookshelves' which show the world what they're reading. It used to be that to know what kind of books someone had on their nightstand you'd have to go to their house, but now all of this information is just a website away. Users can also create lists and collections, and tag, recommend and review titles. When I was doing research for *Print is Dead*, I created a page at Library Thing, listing every book I had read for research. By drilling down on those books, and seeing the books in the collections of users who also had some of my books in their collections, I was turned on to books I would not have heard of otherwise. This is just one of the ways in which the Internet is promoting the discovery of books.

At the same time, online search is allowing worldwide discovery, and putting mountains of information at people's fingertips. Google is the search leader, so much so that 'Google' has become a verb; it's no longer just a website, it's something you do, or have done to you. And readers, interested in either an author, title or genre, are online looking for information. This gives publishers an amazing chance to connect with readers eagerly looking for material to read.

Of course, before writers can be found by Google, their works or their information have to be part of Google's immense online index. For this to happen, publishers need to have vigorous digital archiving projects which not only digitize the works of their authors, but the content of individual titles must be appropriately formatted for online use and tagged with multiple keywords so that it can be found through search. Some of this can be done automatically, given the information that publishers already have in their databases. But for the most part this will involve a new workflow on the part of publishers, on behalf of readers.

Also, publishers need no longer go by the often frustrating classifications that bookstores have long imposed on them, pigeonholing certain books and authors into categories that they think fit. This happened to me when my first novel, *Our Noise*, came out in 1995. It was a paperback original written about Generation X, focusing on the lives of a dozen or so slackers living in a small town in Virginia. One afternoon, after searching for it in a bookstore shortly after it first came out, I couldn't find it anywhere. Finally, as I was leaving the store, the cover caught my eye. Because of my last name, I guess, I had been filed under Latin American interest. No offense to great books like Richard Rodriguez's *Hunger of Memory*, but that's not where my book belonged.

Now that readers have been empowered by Google with the ability to intuitively search by keywords and topics, they no longer have to organize their interests in the same way that a Barnes & Noble shelves their books. But in order for a publisher's books to appear in the search results, there must be some tagging. It'll be a bit of work, but it will bring a large reward.

Digital reading will also help readers share the books or passages that they like with others. That is, if publishers allow them the ability to forward the books to friends with comments or otherwise share their text. And if books don't have this ability, then plenty of other kinds of electronic content will take their place. Just about every Web page you visit that has any appreciable amount of content on it provides a feature where you can email it to a friend. Blogs allow users to subscribe to their sites via RSS feeds, syndicating their content throughout the Internet and the world. And of course what you're doing when you go to a blog is reading, and sharing in an experience (which hundreds if not thousands of others are also sharing). Through the comments section, readers can leave feedback with the writer or else exchange viewpoints with other readers of the blog. This is the kind of discussion that used to happen between friends or family members who have read the same book. Today that kind of exchange is occurring mostly online, in forums and message boards and chat rooms. If publishing can't find a way to tap into this need for discussion, then it's going to find itself and its products increasingly left out of the conversation.

In 1965 Gordon Moore, the co-founder of computer power-house Intel, made an observation that later became widely accepted and known as Moore's Law. He predicted (in effect) that computers would double their speed every eighteen months. And as computers have gotten faster – because of the reduction of chip size and other factors – they've also gotten cheaper. This is why, in the thirty years since personal computers were introduced, they have become both more powerful and less expensive. It has been predicted that Moore's Law will be in effect for at least the next two decades.

With the potential to have perhaps just as profound an effect as Moore's Law on our every day lives, a new theory named Kryder's Law has recently emerged. Based on the work of engineer Mark Kryder, this new law has to do with computer memory. Kryder and his colleagues are working on new ways, and new devices, to store increasing amounts memory.

'Since the introduction of the disk drive in 1956,' wrote Chip Walter in *Scientific American* in 2005, 'the density of informa-tion it can record has swelled from a paltry 2,000 bits to 100 billion bits (gigabits), all crowded in the small spaces of a square inch. That represents a 50-million-fold increase.'

And while memory may seem boring next to speed and price, the larger memory capacity of devices and computers – and disc drives themselves – have led to things like iPods and digital video recorders. All of those billions of downloaded songs, not to mention on-demand TV, couldn't exist without increased and cheaper memory drives.

The combination of Moore's Law and Kryder's Law means that computers and electronic devices will get faster and cheaper, and will be able to hold more of what people are beginning to use their computers and other devices primarily for: consuming entertainment. Even within the past decade we've seen amazing leaps in storage capacity and physical size. When the iPod was first introduced, in 2001, it offered a 5 Gbyte hard drive that allowed you to put '1000 songs in your pocket.' In 2007, Apple was offering an 80 Gbyte iPod (thinner and cheaper than the original model), that allowed users to carry in their pocket up to '20,000 songs, up to 25,000 photos, and up to 100 hours of video.' That's a big pocket.

In the future, disc drives, handheld devices and tablet PCs will be able to hold entire film and music libraries, not just of one director but of every song or film ever produced. What this means to readers is that they will be able – in the same way as iPod owners today – to carry around their entire library of books right in their hip pocket, on an electronic device no bigger than a mass market paperback. In fact, *The New Yorker* sells a portable hard drive that carries over 4,000 issues of the magazine, from February 1925 through April 2006, on a device about the size of a deck of cards. And of course the storage of books and magazines that people already own will be fantastic enough, but consumers will also, through wireless connections and web browsers, be able to purchase content from anywhere in the world and have the text instantly downloaded into their device.

With the combination of these laws – Moore's and Kryder's – electronic reading will finally seem to make sense to most readers. Even early reviews of eBooks, while usually deriding the experience itself, grudgingly admired the fact that a few dozen books could be held on a device at one time, with memory cards allowing for the storage of hundreds more.

'By far the best things about the Reader are its capacity – it can hold about 80 books, or more if you use a memory card – and portability,' wrote Charles McGrath in *The New York Times* in 2006. 'I have crammed mine with both recreational reading and stuff for work, and it all fits in my pocket.'

While comparing the reading experience of physical books to electronic books, it was never a fair fight; books always had the advantage. But with super-fast and inexpensive devices that can hold hundreds if not thousands (not to mention tens of thousands) of titles, digital reading finally has something to offer that print books don't. Suddenly even the most staunch critics of digital reading see that it makes sense.

Portability, searchability, and the fact that you can carry around every book you own at once; these are the real hallmarks of a digital reading experience. It's not about the page versus the screen in a technological grudge match. It's about the screen doing a dozen things the page can't do. What's going to be transformed isn't just the reading of one book, but the ability to read a passage from practically any book that exists, at any time that you want to, as well as the ability to click on hyperlinks, experience multimedia, and add notes and share passages with others. All of this will add up to a paradigm shift not seen in hundreds of years.

'As easel painting deinstitutionalized pictures,' wrote Marshall McLuhan in *The Gutenberg Galaxy: The Making of Typographic Man*, 'so printing broke library monopoly.' Libraries used to be where most people had to read literature or learn about information. But in the future, we will be able to hold in our pocket a library's worth of books.

In his short story 'The Book-Bag,' W. Somerset Maugham writes about a character who lives in fear of being without appropriate reading material close at hand, especially when on trips. 'Since then I have made a point of travelling with the largest sack made for carrying soiled linen and filling it to the brim with books to suit every possible occasion and every mood. It weighs a ton and strong porters reel under its weight. Custom-house officials look at it askance, but recoil from it with consternation when I give them my word that it contains nothing but books. Its inconvenience is that the particular work I suddenly hanker to read is always at the bottom and it is impossible for me to get it.'

As silly as the situation is, it's not terribly different from most people when they go on a long vacation. They'll take three or four books in case they finish one or don't like the other, or if they won't be able to find something else they want to read while they're away. For people hooked on reading, they'd rather be stranded on a desert island with books that they're interested in than stranded anywhere else with books they don't want to read. Future generations, because of digital books and bigger, cheaper computer memory, will be able to carry around every book that they own on a laptop or other portable device. And even if they grow weary of their own collections, they'll be able to wirelessly hook up to an online bookstore and buy a new book, beaming it instantly into their device or computer.

'There is no doubt that, due to the nature of people's music consumption habits, mobility is the major driver behind the latest developments,' wrote David Kusek and Gerd Leonhard in their book *The Future of Music: Manifesto for the Digital Music Revolution*. 'People like to listen to music everywhere – at home or on the road, while waiting, or while socializing. The total support of the mobility paradigm is what drove radio, in the early days, and that's what will drive digital music.'

Books have been more or less mobile since Gutenberg freed them from the chained carrels of the monastery, and it's long been one of the best thing about books: the fact that you could tuck one into a bag or pocket and – if given a few minutes – you could pull it out, start reading, and be transported to another world. Paperbacks, especially, have made books incredibly portable, and most booklovers – like the one in Maugham's story – carry a book around with them at all times. However, unlike the character in Maugham's story, they usually only carry around one book at a time. That, however, is about to end.

Publishers need to learn from the mistakes of the eBooks debacle earlier in the decade, and allow readers the freedom to access their digital libraries wherever and whenever they want. The whole point of electronic reading is that it offers something different from the experience that comes from reading a printed book, so if legally downloaded material is shackled to a certain device or computer – the way that it has been with most modern digital rights management – then consumers won't see any added incentive to this and so adoption of digital reading will continue to falter.

'The thing about an e-book is that it's a social object,' wrote Cory Doctorow in *Forbes Magazine* in 2006. 'It wants to be copied from friend to friend, beamed from a Palm device, pasted into a mailing list. It begs to be converted to witty signatures at the bottom of e-mails. It is so fluid and intangible that it can spread itself over your whole life.'

That fluidity is the key to the future of publishing, giving readers the ability to share the books and texts that they love with family and friends. In a virtual world where email and digital photographs have become the way that people communicate and share experiences, readers will want to also share their digital books. As people become geographically spread out, and are always online, the age of handing a friend a physical book will disappear but the impulse will remain. Publishers need to give consumers this ability.

'We must get used to the idea of this free flow of information being the norm rather than the exception,' wrote Kusek and Leonhard in *The Future of Music*. 'We must realize how the power of online networking contributes to the fulfillment of the needs and desires of the digital kids, especially once it intertwines with offline, "real-life" events and experiences.'

It won't just be about the sharing of books (one consumer lending or even giving the work to someone else). It will also be about the accessing of one book from multiple points by the same consumer. Someone who has purchased an electronic book should be able to access and consume that material nearly anywhere and in nearly anyway that they want to. Instead of sitting on a bookshelf at home, purchased text will instead live on a virtual bookshelf on a remote server to be accessed from any computer or handheld device that can go online to retrieve that text.

This would mean enormous convenience to readers, and could have come in handy for me a few years ago. Right before I took a Christmas vacation I started a new book, Kazuo Ishiguro's *The Unconsoled*. I'd only read about a dozen or so pages before getting on a flight from New York to California – where I'd be spending the next couple of weeks – but I was eager to get into the long novel, one I'd been meaning to read for years. But when I got to the airport I was extremely upset to discover that I'd left the book sitting beside my bed instead of packing it in my carry-on bag. I'd been looking forward to reading it on the flight, as well as while I was in California. Instead, it was sitting at home in my empty apartment.

As soon as I touched down in Los Angeles I had a relative drive me to the nearest bookstore, where I promptly bought another copy of the book, not able to wait until I got back to New York to continue reading it. Now, while publishers may love this story (since I bought the same book twice), as a reader I found it immensely frustrating. What if I could have, once I'd discovered at the airport that I'd left the book at home, downloaded another copy of it onto my laptop while waiting to board the flight? After all, I'd already legally bought the text. Why shouldn't I have access to it anywhere I went? And if this kind of experience and functionality were available, how many more books would consumers buy? There can't be much doubt that readers would buy more books rather than fewer if those books gave them more options and ways to explore them.

In the end, *The Unconsoled* ended up being one of my all-time favorite books. As soon as I got home I mailed my first copy to a friend of a friend in London who I thought would enjoy it, and I then put my second copy on a bookshelf. Since then, I've thought about Ishiguro's strange world hundreds of times, rolling the scenes and images over and over again in my head the same way I've closed my eyes and tried to remember a brilliant painting I saw years ago. In a delightfully low-tech way, *The Unconsoled* is always with me because I keep the story and its world with me at all times (in my mind, that is). In terms of the paperback I bought in LA, I haven't touched it since I finished reading it.

My experience with the Ishiguro novel also illustrates that books are actually sometimes troublesome nuisances that get in the way of our experiencing the story. Often I've been in bed with a large, *Gravity's Rainbow*-sized hardback book that I could barely hold in one hand. And try casually popping Robert Caro's biography of Robert Moses, *The Power Broker*, into your bag in order to do some lunchtime reading. The thing is over 1,300 pages long, and weighs more than three pounds. Reading it is a delight, but having to carry it around is not.

The notion that the story and the emotions behind it are what make a book special has been around for decades. In the 1980s, Anthony Burgess declared that 'a novel ought to leave in the reader's mind a sort of philosophical residue.' More than twenty years later, author and marketing expert Seth Godin wrote on his blog in 2006 that 'a non-fiction book is a souvenir, just a vessel for the ideas themselves.'

Both Burgess and Godin are onto something. The 'residue' is what we remember when we think or talk about the book after we've read it. As much as we like to think that they belong on shelves, the true place for books is inside our heads. Books are just husks, 'souvenirs' if not trophies ('Look how smart I am to own *Foucault's Pendulum*!'), and yet what has an effect on us is the story, the prose, the characters and plot. What affects us least is the paper. The same goes for music. When you buy a copy of *Abbey Road*, and you find yourself humming the bassline to 'Come Together' for a fortnight, the music has left its 'residue' in your mind in the form of a melody. The LP, CD or even the MP3 lurking on your iPod's hard drive is just a souvenir.

'At another level, absorbing literature in a variety of ways – onscreen, on paper, on tape – helps to dispel the false equation between text and book,' wrote Jacob Weisberg in *The New York Times* in 2000. 'Powerful associations from childhood – the smell of must, the flashlight under the covers –

have bred generations that think of themselves as book lovers rather than story hounds or prose fanciers.'

It's the presence of this 'false equation' that, if not changed, will doom the world of print to a ghetto of tweedy collectors and literary snobs. Maureen Corrigan wrote, in her 2005 book *Leave Me Alone, I'm Reading: Finding and Losing Myself in Books*, that 'what we readers do each time we open a book is to set off on a search for authenticity.' No, when you open a book what you're doing is prying open a physical object. The most important thing to be opened during the reading of a book is a person's mind, not two slabs of cardboard. The 'search for authenticity' that Corrigan is writing about can take many forms and lead in many directions, of which books are just one. Today's challenge for books is to stop becoming the road not taken.

Even when critics and academics reluctantly acknowledge the advantages of digital delivery, they *still* prefer paper. Author Edward Tenner wrote an essay about this entitled 'The Prestigious Inconvenience of Print,' that appeared in *The Chronicle of Higher Education* in 2007. Tenner's thesis was that, even though digital media provides an interactive experience and is (mostly) all-around more efficient, the physical medium of print still contains within it a 'prestigious inconvenience' which people are willing to put up with. Tenner's real point is basically the chant of the pigs at the end of *Animal Farm*, which I'd paraphrase as 'Digital good, print better!' For instance, he mentions that while most businessmen send and receive voluminous amounts of email, 'their most important sentiments are likely to be expressed as handwritten notes – one of the reasons for the luxury fountain-pen industry's niche in the digital age.' In this Tenner is correct, especially in his use of the word 'niche.' Fountain pens are a rarity these days, having been replaced by the computer, Blackberry, and handheld devices with keyboards such as cell phones and

Palm Pilots. This is the argument of the 'print is dead' debate; not that print will become extinct, but that it will instead become a niche product and specialized interest.

Tenner even hovers around the idea of print being dead, writing that:

> just as luxury watches remain in demand while most people carry cell phones that give the time with virtually observatory-standard accuracy, the Web will never destroy older media because their technical difficulties and risks help create glamour and interest. At the same time, however, the Web does nibble at their base, creating new challenges for writers, musicians, and other members of the media.

First of all, I would say that the Web is doing a lot more than just 'nibbling' at older media. The thousands of people who have lost their jobs because they worked at magazines or newspapers that went out of business due to lack of interest and online competition would probably say they feel swallowed whole and not just nibbled. And here Tenner makes a point he probably doesn't want to make, pointing out that one day (perhaps soon) a printed book in a digital world will seem as quaint and as antiquated as a fountain pen feels today.

One of the biggest questions that so far has remained unanswered during the initial stages of literature's transformation from the printed page to the digital screen is what exactly electronic books will cost. What will be the price of these works that exist only as computer files? What will publishers charge, and what will consumers be willing to pay?

In the first six years of eBooks' existence – an interesting time because print books existed alongside electronic books – consumers almost always demanded a much lower price for a digital edition due to the perceived lack of value since 'all they were getting' for their money was an electronic file and not a physical object. Later generations may be willing to pay a premium for the convenience of a digital file and all of the mobility that comes with it, especially if it's stripped of digital rights management and can truly be read on any device.

Consumers wrongly think that the bulk of a book's price is for the printing and shipping. They don't realize that there are myriad costs involved that a publisher is trying to recoup under the umbrella of that price, among them: author advance, future royalties, editing, copy-editing, marketing and publicity costs, and of course basic office operating expenses such as paying for office space, staff salaries and storage costs. But the majority of consumers think that the bulk of a hardback's $27.95 list price should go away when they buy an electronic edition. The publisher isn't paying for printing all of those pages, and they don't have to produce that colorful dust jacket. So why shouldn't the electronic edition of that $27.95 hardback cost half that, or even less?

In the comments section of a *Business Week Online* article from 2005 entitled 'Curling Up With a Good E-Book,' which focused on Sony's new eBook device but also discussed the market in general, two consumers weighed in on the subject. One wrote, 'Sony needs to be very careful to price the books right to avoid piracy. Some [*sic*] $1 or $2 dollars is enough considering there is no cost in duplicating the work. $1 is actually more than the author receives for a paper novel. $1 for the author and $1 for the publisher should be more than adequate.' Another one wrote, 'The problem I see is that I think consumers will balk at the prices close to mass-market paperbacks. $7–$15 is too expensive for an electronic copy of a book that needs the Sony reader to read. Prices should be closer to $2 a book.' This is the kind of thinking that the publishing industry needs to contend with if it's going to segue into a world of digital reading.

To combat this perception, publishers need to emphasize the fact that a book is an experience like going to the movies. When you go to the movies, all you leave with is your ticket stub. What you're paying for is the experience of living with that movie for two hours. The ten or fifteen dollars of the ticket price goes mostly to the studio producing the film. Books must be similarly produced, with writers and agents and editors all working to produce a finished product.

The danger is that the value placed on content now will likely remain (and indeed, haunt the industry) for a long time. A delicate balance needs to be struck so that consumers feel they're getting value for money, and so that the publishing industry makes enough to stay in business.

The temptation is to price electronic editions temporarily low to kick-start the electronic reading revolution or to move eBook-related merchandise such as dedicated reading devices. But once prices are set at a low level for content it will be very difficult, if not impossible, to raise them. After all, consumers are not going to be thrilled if they get used to paying one price for an electronic edition of a book, only to find the price for the same product double in a few years.

This is also where the digital adeptness of Generation Download becomes a liability. They've grown up on downloading songs or entire records from file-sharing websites or blogs without paying for them. There's often the perception that such an act is an innocent thing, that one or two downloads won't hurt the recording industry (or the band whose music they're downloading). Or else they feel that if it was an illegal act – like shoplifting a CD in a store – it would be harder to pull off. And despite various intense efforts from the recording industry – including lawsuits and public relations campaigns – they have been only marginally successful in changing people's minds about the value of content. In fact, many efforts (such as suing file sharers) have backfired, and only added to the perception that the record companies and the industry itself are greedy, out of touch and ripe for a fall.

In terms of reading material, there's already a plethora of free content available; blogs, websites and wikis all offer hundreds of thousands of words everyday for free to anyone with an Internet connection. Kids are just used to reading stuff online. And when sites that have previously offered all their content for free suddenly put some content behind a paid barrier (like *The New York Times* did a few years ago with its columnists), many simply go elsewhere.

What's difficult is both making that transition (from free to paid content), as well as instilling in the mind of the users that they are indeed getting their money's worth for a virtual 'product.' And kids who surf the Web all day aren't used to coming up against the boundaries of 'paid' anything; they're use to file-sharing, song swapping, tinkering with open source code, website scooping, and instantly downloading anything they want (records, TV shows and even movies). When they finally get to the point where digital books are on their radar, they're going to zero in on them the same way they would any other form of content, and they may expect it for free. In fact – unless they're educated about the value of content – this generation will likely think the same thing about books as they did about records in the era of Napster: 'So I downloaded someone's scanned copy of *Franny and Zooey* from a website and read it on my laptop... where's the harm in that?'

Nonetheless, the electronic future is not as bleak as it seems. The tremendous success of iTunes, with its billions of downloads, has shown that – with the right software interface and device, paired with the right price point – people are more than willing to pay for legally downloaded content. Indeed, the more comfortable with downloading they get, the more they replace their real content for their virtual content (which is why sales of CDs continue to plummet while iTunes' numbers continue to rise). Piracy is still an issue, but it shows that if you offer consumers legal content for a fair price, they'll buy it.

While it took the recording industry years to figure out a model that works – and many would claim that they still haven't, and that the Internet has begun the inevitable decline of the music business – publishing is still in the preliminary stages of figuring out what to charge readers for electronic works, and what to pay authors for the sale of these virtual files. Here's where publishing can learn from the mistakes of other industries, and leapfrog some of the growing pains from going digital.

'I don't think it's practical to charge for copies of electronic works,' wrote Cory Doctorow in an article in *Forbes* in 2006. 'Bits aren't ever going to get harder to copy. So we'll have to figure out how to charge for something else. That's not to say you can't charge for a copy-able bit, but you sure can't force a reader to pay for access to information anymore.'

What the price points will ultimately be for literary content, or how the business model works in general, remains to be seen. But something needs to occur so that we establish solid notions – in the minds of consumers – of the value of downloaded or online-accessible literary content. Until that happens, publishers leave themselves open to confusion in the marketplace, continued resistance against widespread digital adoption, and perhaps piracy and theft.

Electronic reading and digital delivery is not just a new way of doing an old thing – issuing books one way instead of another – it is instead an entirely new way of doing business. As revolutionary as reading will be in a digital future, so too must be the accompanying business model. If not, consumers will reject paid content and surf the Web until they find something that they can read for free.

10

will books disappear?

Wʜɪʟᴇ ᴛʜᴇ relevance and popularity of printed material (such as books, magazines and newspapers) will get smaller and smaller over the next few decades due to digital reading, books themselves will never entirely go away. Instead, they will be sought out by collectors, those who want to hold and touch pages, covers and dust jackets. And books will always have a place in millions of homes across the country, but they will become rare as printers go out of business and warehouses gather dust. While some people will continue to do the majority of their reading in an ink-on-paper format, their numbers will increasingly dwindle and we'll see a reversal of the situation we see today where the majority of news and fiction is printed on paper, while only a small proportion is available just in digital format. In the future, most news and communication will be transmitted digitally, while only a fraction is conveyed via print.

In the face of digital reading's dominance, books will turn into a specialized taste, an art form. In their 2004 book *The Fall of Advertising and The Rise of PR*, Al Ries and Laura Ries discuss how candles became an art form after electricity effectively removed them as a necessity for creating and sustaining indoor light. What happened after Edison invented the lightblub, wrote the Rieses, could be described as 'the fall of the candle and the rise of the lightbulb.' They continue:

> Yet every night all over America millions of candles are burning. No romantic dinner is complete without candles on the table. Individual candles are sold for $20 or $30 each, much more than a lightbulb. Unlike an electric bulb, the value of a candle has no relationship to its light output. Like the fireplace and the sailing ship, the candle has lost its function and turned into art.

The same thing will happen to books. Once people not only grow accustomed to getting and consuming their chosen information through a digital interface, but begin to expect and then demand it in that format, books will – after they have lost their general utility – be retained in many people's lives as works of art. People will still want and read books in the immediate decades and following century (the same way that candles have continued to be a common item in households for the hundred years since Edison invented the lightbulb), but they will be increasingly seen as anachronistic, antiques. So for all the dire predictions about print being dead or extinct, it will never go away completely. Printed copies of great novels like *The End of the Affair* and *Of Human Bondage* will still exist, but they will be more like collectors' editions rather than the primary way to get the material.

In addition to classics, a small portion of physical books will continue to be printed for the specialized audience who will want to read them. In this, the general market for printed books in the future will be very similar to the large print market of today. Large-print editions cater to a small and specialized audience. Only bestsellers and genre fiction are made available in these editions – brand name authors like Dan Brown, Danielle Steel and James Patterson – and the total number of large-print editions is only a small percentage when compared to the overall number of books published annually. A consumer who reads large print editions is incredibly limited in the books available for him or her to read.

In a few decades, fans of printed books will be in a similar situation: as more novels and non-fiction titles are distributed solely through digital means, they will find it harder to obtain a wide variety – in print, that is – of things they want to read. Of course, if they're fans of Dan Brown, Danielle Steel and James Patterson, they will be fine. But because more and more new material will be available only as electronic editions, it will lead to more reading through digital means. And the more people who get accustomed to it every year, and take it on as their primary way of reading, the less demand there will be for printed books and thus the availability of physical books will continue to shrink. Readers have shown that they care about content, and will follow it anywhere. The massive success of *The Da Vinci Code* shows that what grabs people's interest is the story. If today that novel were only printed on cocktail napkins, people would no doubt continue to buy it.

Where people buy books today is mostly at huge bookstores like Barnes & Noble and Borders. These retail chains, along with the rise in online shopping and the discounts and service that Amazon and others can provide, have been steadily throttling the world of small, neighborhood bookstores for years.

'Independent bookstores, of course, have been under siege for nearly two decades by the megachains and the Web retailers,' wrote Julie Bosman in *The New York Times* in 2007, 'and have been steadily dropping away, one by one. Now, though, the battle is reaching some of the last redoubts.'

As the small bookstores go under, the only alternative in most towns is either a superstore like Barnes & Noble or else any number of even bigger stores like Wal-Mart, Target and Costco, all of which have book sections that sell the latest bestsellers. In addition, non-traditional venues such as Starbucks have begun to sell books, albeit only a limited selection.

In a future populated more with digital downloads than with physical books, there will be a resurgence of the independent bookstores that have, until now, almost ceased to exist. Why? Because these small bookstores can do things their bigger competitors cannot.

'Yet while closings get attention, bookstores are opening, too. In the last several months, owners have cut ribbons on five new independent bookstores across [New York] city,' wrote Mike Peed in *The New York Sun* in 2007. 'Taken together, these nascent stores suggest that, even in an era of ever expanding conglomerates and one-click Internet shopping, where used books frequently cost less than their postage, independent and niche bookstores just might remain relevant.'

They will do so by refashioning themselves as something more than just a bookstore, either choosing a particular specialty to focus on (such as art books) or else becoming more of a neighborhood hangout, a book 'lounge' featuring a café with food and even liquor. These stores will resemble more a local gift shop rather than the book superstores of today, or even the bookstores of yesterday that sold just books. In fact, many stores are getting rid of the books, turning what used to be mazes of bookshelves into community gathering places.

Writing about the transformation of bookstores in an Internet age, the Associated Press documented the change of a Northern California store named BookBeat, and the steps its owner – Gary Kleiman – had taken in order to stay in business. 'He tore down shelves and in their place put tables and chairs and a small stage for live performances. He started offering free wireless Internet access. And to help convince people to take advantage of it all he got a beer and wine license.' But did it work? 'While he's still selling about the same number of books as he used to, new books are selling better. And his store has a lot more customers – eating, drinking and listening to music – than he did before.'

The key to saving the independent bookstore – especially in an era of rising digital delivery and consumption – is the same thing that's starting to save journalism: think local. Don't try to compete with Amazon or Barnes & Noble, blogs and MySpace. Instead, do what they *can't* do. Of course, for some stores, even this strategy won't help.

Self-described bibliophiles Tom Wayne and Will Leathem owners of the Kansas City bookstore Prospero's Books, decided in 2007 to burn their inventory of 50,000 titles after they could not sell or even give the books away. In a scene that really can't help but sound like it's from *Fahrenheit 451*, over Memorial Day weekend the two men dragged a few boxes of books to the sidewalk in front of their used bookstore, showered them with lighter fluid, and then set the whole thing ablaze.

But this was no Nazi bonfire. It was more like the Buddhists in Vietnam in the early 1960s who committed suicide by setting themselves on fire as political protest. Neither Wayne nor Leathem felt that books should be burned or destroyed – on the contrary, they're both ardent booklovers – but they did this to attract attention to the fact that books were, well, no longer receiving any attention.

As Dan Barry reported in *The New York Times* shortly after the incident, 'The men say they tried to give away books in bulk that were either not selling or in overabundance – to no avail. When a friend was sent to state prison, for example, they tried to donate books to the correctional system, but were denied. When they donated books to a local fund-raising event, some well-meaning person bought up most of those books and left them at the Prospero's doorstep.'

One of the sad ironies of this is that whenever I speak about the 'print is dead' debate, I always point out that it doesn't mean things like what happens in Bradbury's novel, stressing that the advocates of digital reading are not the exterminators of the printed word. And yet here we have booksellers – not technologists – who are lighting the match and turning novels and non-fiction into ash.

Also, remember that, in *Fahrenheit 451*, books are never explicitly banned outright by the government; they only begin to be burned when the public apathy for books grows so large that the government figures no one will miss them and for the most part, no one does). Burning books was only the government's reaction the public's reaction. In Kansas City, we might have seen the sad first glimpse of this happening in real life.

n many ways, the publishing industry has already stared down a previous challenge to its existence in the late 1990s. When eBooks were first introduced, people predicted that they would spell the beginning of the end for publishers. The argument was that writers could go directly to their audiences, bypassing the literary establishment who were already beginning to be talked of as dinosaurs. The newly ubiquitous Web was going to empower writers in a way that would shift the balance of power away from publishers and towards them. Sound familiar? Throw the word 'wiki' into the mix and it resembles very much the debate today.

'Authors frequently grouse about their conventional publishers, usually bemoaning the quality of the editing, the amount of publicity, or the size of their paychecks,' wrote David Kirkpatrick in *The New York Times* in 2000. 'Some authors now hire their own editors and publicists. Plenty of companies already offer the rest of the services publishers provide – from book design to printing and distribution.'

One of the first authors to really challenge the publishing industry in light of all of the opportunities the Internet pro-

vides, was also one of the most successful: horror writer Ste
phen King. King, whose short story eBook 'Riding the Bullet
was the first (and remains the only) real eBook success, started
an experiment in 2000 that would digitally provide content –
for a fee – directly to his millions of fans and readers. The new
sent shockwaves throughout the book industry, and King him
self joked that his direct-to-consumer novel, entitled *The Plant*
(which he was going to write and release in installments), wa
poised to 'become big-publishing's worst nightmare.' For a
while, people believed him.

What finally happened ended up being a nightmare for Ste
phen King (and a vindication for publishers). He began by
charging readers $1 per installment of *The Plant*. Readers paid
on the honor system, and King promised to keep posting the
installments on his website if at least 75% of the people who
downloaded the story paid the dollar. Over time, readership
declined, so King raised the price. People kept downloading
(though at much smaller numbers) and King kept raising the
price. That is, until he finally gave up and ended the project
without ending the story.

After all of the hype, and all the hand-wringing and hubris
that surrounded its launch, *The Plant* finally withered on the
vine. The popular technology website Slashdot said of the epi
sode in 2000, 'What King's adventure demonstrates is that the
Net is a powerful new tool for selling books rather than a tech
nology that replaces them.'

Thanks to King's experiment, instead of proving that pub
lishers weren't needed, it proved the opposite. People could
see that, even with a sure-thing like King, it wasn't as easy as it
looked, and that perhaps – even in a world of digital conve
niences – publishers still had a role to play.

Another technology in the late 1990s that held out the hope
of changing things for writers, and leading to the death of the
big-time publishers, was print-on-demand (also known as

OD). This new technology was made possible by machines that could print one book at a time, reproducing color illustrations and even dust jackets. Of course, for decades now a printing press had the ability to produce only a handful of books, but to do so – because of the prices involved – would have been commercial suicide. What the print-on-demand machines allowed was the printing of a handful of copies for only a handful of dollars.

Shortly after the POD machines were invented, companies sprang up offering to produce, and in some cases promote and edit – for a price, of course – the books of would-be authors. There were several levels of service available, and many of them – including writing flap copy and designing a cover – were the same things an author would get at a major publishing house. Many journalists and pundits came right out and said that if all publishers do is get you an ISBN and put your words between covers, then here were companies who would do just that (sometimes giving even a larger share of the profits than the major publishers; that is, if the book ever sold any copies).

At the time, Xerox (one of the makers of print-on-demand technology) produced a commercial in which a young student stands up in class and rebukes his stodgy old professor, who is in the middle of lecturing the class about how they will most likely never grow up to be published authors. The obstreperous student states that – because of all this great new technology – one day *everyone* will be published.

True, perhaps, but only to a point. The Xerox machines do indeed allow someone to produce a professional-looking book. However, as author and marketing expert Seth Godin wrote on his blog in 2006:

Publishing a book is not the same as printing a book. Publishing is about marketing and sales and distribution and risk. If you don't want to be in that business, don't! Print-ing a book is trivially easy. Don't let anyone tell you it's not. You'll find plenty of printers who can match the look and feel of the bestselling book of your choice for just a few dollars a copy. That's not the hard part.

The hard part is of course getting someone to want to read i instead of all the other thousands of books that are publishe each year by established publishers (not to mention the hun dreds of thousands that are produced via print-on-deman technology).

Of course, what most print-on-demand companies amoun to is really just a new spin on the old vanity press companie that have always been around. So a company makes you book, complete with a nice cover and even an ISBN, and i shows up on Amazon. But can these authors really conside themselves *published*? POD publishing tends to lead to mor questions than answers.

'Is print-on-demand publishing purgatory or a legitimat venture?' wrote Elaura Niles in her 2005 book *Some Writer Deserve to Starve!: 31 Brutal Truths About the Publishing Indus try*. 'This is an area of publishing that's still defining itself.'

Now, more than five years after Xerox's commercial firs aired, people no longer speak of print-on-demand as being the death of publishers. Instead, POD technology is looked a as something that regular publishing companies can use to quickly reproduce titles and cut the cost of warehousing hun dreds of thousands of books. Many mainstream publisher today are selling their backlist titles as print-on-demand, avail able through any number of online retailers. Consumers who purchase them over the Internet never know the difference. This allows publishers to offer many more books than they could keep in their warehouses. For authors it means an incre mental income and that their books will never go out of print.

Even a demand of just a dozen or so copies a year is enough for a publisher to keep it in their system as a POD title.

A few of the original print-on-demand companies – such as Random House's venture Xlibris – are still around today, as are a number of newer, hipper companies (most notably Lulu and Blurb), offering authors inexpensive print-on-demand services. In a matter of only a few days, aspiring authors can be proudly clutching their books. With this kind of speed and ease of use, sometimes it's easy to see why so many people have rushed to the conclusion that all of this technology might indeed spell the end of the publisher. Instead, the various technologies are new chances and opportunities for publishers, authors and readers.

'For a century we have winnowed out all but the best-sellers to make the most efficient use of costly shelf space, screens, channels, and attention,' wrote Chris Anderson in *The Long Tail*. 'Now, in a new era of networked consumers and digital everything, the economics of such distribution are changing radically as the Internet absorbs each industry it touches, becoming store, theater, and broadcaster at a fraction of the traditional cost.'

As Anderson showed in his influential book, even selling just dozens of copies a year of a POD title can add up to an immense profit over time (if there are thousands of them spread across a publisher's backlist). This is especially true if those books were out of print or unavailable; revenues are practically found money. But now we're starting to see POD machines accessible to the everyday consumer. One of the most exciting and promising of these machines is named Espresso.

Described as an 'ATM for books,' the Espresso is basically a vending machine for printed material, instantly producing a custom-made book from digital files. As the dominance of the huge bricks-and-mortar stores gives way to machines like

Espresso, print-on-demand technology could fill an extremely useful niche, turning any location into a bookstore. Take the typical airport bookstore. Why do they carry only an extremely narrow and commercial selection? Because of space. If an airport bookstore had a few Espressos, consumers could choose from millions of titles instead of from just a few hundred. And it wouldn't have to be just bookstores that have the machines. Since the Espresso and similar machines take up a small amount of space, any location could become a bookstore. They could stand alongside soda and snack vending machines in high schools or at malls.

'Buying a book could become as easy as buying a pack of gum,' wrote Emily Maltby in *Fortune Small Business Magazine* in 2006. 'After several years in development, the Espresso – a $50,000 vending machine with a conceivably infinite library – is nearly consumer-ready and will debut in ten to 25 libraries and bookstores in 2007.'

So while publishers are far from being run out of business by either electronic books or print-on-demand technology, that's not to say that certain departments of publishers won't be greatly affected by a digital future. In the same way that many authors will not survive the transition to electronic books, many areas of publishing will also have a hard time adapting to literature's new digital landscape. The departments most affected will be those who deal with either the manufacturing, storing or moving of those printed products that will one day go away. But the departments whose job it is to create the words that once used to fit onto printed pages – soon to be delivered through cable modems and wirelessly through the air – will operate in the future much like they do today: finding and grooming talent, acquiring and shaping stories, advising and nurturing young authors.

Because, while what was previously known as a book will no longer need the 'box' of pages and binding, the knowledge contained therein will still need to be found, edited and marketed.

The future of the book debate then becomes merely a question of altered consumption. The only thing that goes away when we look at the future of reading is the shell of the book, the husk of the actual physical pages; the content will of course remain. In the end, printing plants have much more to fear from a digital future than publishers do.

And yet, some people continue to make the argument that publishers will soon be irrelevant. After all, if print is to be delivered digitally, why would writers still need a company to act as a middleman between them and their readers? Won't publishers just get in the way? Won't novels end up being like blogs, with writers using free software to instantaneously connect with their audience?

These are all interesting questions, and to try to answer them I've come up with the following list:

Five reasons publishers will still exist in a digital age

#1 Find talent
With millions online, finding anything worth consuming is getting more difficult.
In the summer of 2006, MySpace registered its 100 millionth user. Meanwhile, videos are being uploaded to YouTube so fast you could never watch them all, even if you quit your job and stayed home to try to watch them all. And the popular blog search engine Technorati proclaims on its home page that it can offer searches of 'zillions' of content-filled web pages. (By the way, 'zillion' isn't a real number; I looked it up.) There's now *so much* content out there that it's not only impossible to try to consume it all, but it's getting increasingly

difficult to even know where to start. What publishers will continue to do – as they have done in the past – is to act as talent scouts to find worthwhile content, and then help to bring it to the surface. Because, as Technorati says of the photos, videos and blogs found in its index, 'Some of them *have* to be good,' which is another way of saying that most of them are *bad*. With so much content already out there, and more being produced each day, publishers will fill an important need and perform a valuable service (for writers and readers alike) by reaching into the digital slush pile and pulling out the pearls.

#2 Support talent

The Internet is great for making an initial splash, but not for turning that splash into a career.

Online infamy is easy to come by, but turning even positive Internet exposure into something that lasts longer than Warhol's 'fifteen minutes' of fame is difficult. The 'guitar' video on YouTube mentioned in Chapter 5 has been viewed, as of mid-2007, over 21 million times. That's an amazing feat, but it's just an exercise if it's not in aid of anything. The young man who made it doesn't have a CD to sell, or even a website on which he can sell ads (for a while, no one even knew who he was). It's fun, but it's not a business model. On the other hand, when the band OK Go became YouTube darlings with their treadmill video for 'Here We Go Again,' they were supporting a record that was funded and released by a major label. The band may have paid for the video themselves, but the record the video was made to promote was paid for by the advance their label gave them. So while it's sometimes too easy to get an audience online, that exposure is only really useful if it's in support of something that users can interact with apart from the vehicle that brought the initial exposure.

#3 Edit talent

Even geniuses need editors.

A skilled editor proves invaluable advice to even the most gifted writer; remember that F. Scott Fitzgerald wanted to name his classic novel *The Great Gatsby*, among other things, *Trimalchio in West Egg*. Yes, it would have been just as good a novel no matter what it was named, but its commercial prospects would have been much more narrow had it been named *Trimalchio in West Egg*, and it might not have survived long enough for its reputation to rise spectacularly after the author's death. And Maxwell Perkins did more than just advise Fitzgerald on the name of the book; he also shaped the author's very career (not to mention that he discovered and edited half a dozen other great American writers, among them Ernest Hemingway and Thomas Wolfe). Without editors, and with writers keeping blogs that they update several times a day, the onus is then placed on the reader to filter and parse the text. 'But of course readers don't want to become editors,' writes Sean Wilsey in *Time*. 'What they want, what I want, is for what I'm reading to have already undergone the sort of editing that allows reading to be an intimate, thoroughly immersive, deeply pleasurable activity.' And for a generation who learned how to write on computers, and have spent most of their youths texting and instant messaging each other, editors will be sorely needed to translate all those emoticons and LOLs to real words. Without editors, books or electronic texts will simply be blogs in a different package.

#4 Expose and market talent

As more authors are discovered online, more authors are promoted online.

Now that the Web has given anyone with a computer global access to information, it's quickly becoming the tool of choice for researching and finding things. In terms of publishing, consumers use the Internet not only to buy books they already know they want, but also to discover new authors and titles. Because of this, traditional methods of marketing books – such as placing ads in a newspaper or magazine – are going away. Using the power of the Internet, publishers will do numerous things to expose and market writers to online communities, including creating banner ads, interactive websites and blogs, as well as performing outreach to bloggers and Internet reading groups. Publishers will also produce video trailers, or videos featuring the authors themselves, which can then be uploaded to YouTube and traded around the Web. Publishers will also maintain email addresses of fans, using them for marketing purposes by sending out professionally designed e-cards and newsletters. Some web-savvy authors can of course handle this themselves – and some do – but for the most part, writers will be most happy (and probably already expect) their publishers to handle all of this work for them.

#5 Pay talent

The Internet creates communities, but it doesn't pay them.
While immensely popular websites like Boing Boing have enough traffic to generate substantial revenue through advertising, most blogs and websites – in addition to most exposure found online – are not money-making ventures. Again, the 'guitar' video that has been seen on YouTube over twenty millions times, while reaching an immense audience, hasn't paid any revenue to its young creator. What publishers will continue to do is sell the works of artists in the marketplace, and then pay royalties on those sales (no matter what those sales look like in a digital world, whether they're for the complete text or chunks of it). And while there are websites, such as PayPal, that writers and musicians are currently using to directly sell their work to consumers, as Stephen King's experiment with *The Plant* showed, most writers don't want to have

o handle the financial side of things. They'd much rather pend their time writing. This is where publishers will continue o come in handy.

As I'm sure you noticed, what all of the above examples have n common is that they have to do with talent. Also, I didn't call them *writers*. Because why limit ourselves? John Sayles is known for directing movies, but he also writes books. Vice versa for Paul Auster. Leonard Cohen got his start writing novels, but everyone knows him today as a singer-songwriter. We want to be involved with talented people, and we don't want to limit ourselves to a particular format. What if John Lennon had begun his career as a writer instead of as a musician (instead of writing books only after he became a Beatle), and he walked into his publisher's office one day and picked up an acoustic guitar and played 'A Day in the Life'? That publisher would be an idiot for showing him the door, telling Lennon on the way out, 'Sorry, son, we're in the book business.'

Simon & Schuster in 2007 had a megasuccess with *The Secret*, a self-help motivational program released as both a book and a DVD. So what is it, something you read or something you watch? Or rather the question should be: What does it matter? The idea is key, and it's all about however the idea happens to reach you. In the same way that Soderbergh released *Bubble* in multiple formats on the same day, publishing will come up with new models that allow for consumers to make their own decisions about how they choose to consume content. Because, finally, content *really is* king.

This was something that was said all the time back at the introduction of eBooks in 1999, but I really don't think that back then anybody believed it. What *was* king, at the time, was technology. It was all about the interface, and not the user; it was even less about the content. Content, in many ways, was an afterthought, if not an excuse. It was the 'MacGuffin' of Hitchcock's films, the incidental nail on which to hang the plot that could, in the end, turn out to be anything. It didn't matter; the adventure was the point. And yet publishing in the future truly will have to be about the content because, if the physical product goes away, the only thing that will be left is content.

In an essay published in 2007 in *The Guardian* entitled 'Fail Better,' Zadie Smith wrote about the quest by novelists to write great books. After a lengthy discussion of a novel's various components, she then got to the heart of the matter, which is basically the effect that novels have on us:

> A great novel is the intimation of a metaphysical event you can never know, no matter how long you live, no matter how many people you love: the experience of the world through a consciousness other than your own. And I don't care if that consciousness chooses to spend its time in drawing rooms or in internet networks; I don't care if it uses a corner of a Dorito as its hero, or the charming eldest daughter of a bourgeois family; I don't care if it refuses to use the letter e or crosses five continents and two thousand pages. What unites great novels is the individual manner in which they articulate experience and force us to be attentive, waking us from the sleepwalk of our lives.

What Smith was getting at (the grand effect that novels have on us) is what they do to our minds and our souls. Nowhere in her essay did she mention what novels do to our *fingertips*. The method through which anyone reads and absorbs a great novel is just about the least important aspect of the process.

'It is only today that industries have become aware of the various kinds of business in which they are engaged,' wrote

Marshall McLuhan in *Understanding Media*. 'When IBM dis-covered that it was not in the business of making equipment or business machines, but that it was in the business of pro-cessing information, then it began to navigate with clear vision.'

Publishing needs to come to a similar conclusion, realizing that it's not in the book business, but instead that it traffics in ideas, information and stories. Just as video and DVDs proved to the movie business that their trade had little or anything to do with film itself, so too will digital reading prove to publish-ers that cardboard and pulp are merely the passing adjuncts to its most important processes. If we can begin to grow in our minds the idea of words being special, and realize that books are just paper, then our clear vision awaits us along with all the benefits that will come from opening our eyes.

afterword

IF YOU MAKE UP a list of movies that offer dystopian visions of the future, you'll find that it's much longer than the list of utopian ones. From art films like *Blade Runner* and *Brazil*, to the popcorn fantasies of *Total Recall* and *The Matrix*, the vision of life on a future Earth – in most movies – is of a dark and brutal world where the only color comes from either advertising or a person's own dreams. Even Steven Spielberg's two sci-fi films of the past decade, *Minority Report* and *A.I.*, portrayed a future that was a mostly cold and dreary place. True, each of these had their source material elsewhere: *Minority Report* was based on a Philip K. Dick short story, and *A.I.* was a project that Stanley Kubrick worked on for years. But the fact that Spielberg chose not to sunny-up the material with adorable aliens *à la E.T.* shows that even he must have a pessimistic streak inside of him somewhere, an impulse that thinks things are getting worse and not better. Conversely, I really can't think of one film that stands as a somewhat idealized portrait of the future, a place where things aren't so bad that a person today wouldn't mind closing their eyes and waking up there.

I've chosen examples from cinema purely because they seem to stand out in the minds of the general public more than books. Orwell's *Nineteen Eighty-Four* is the obvious exception, not to mention that it is the rare example of a work where people think of the book rather than the movie; try doing that with *A Clockwork Orange* (another dystopian vision, by the way). But even in Orwell, it's the ideas that have staying power. The notion that Big Brother is always watching us on small silver screens is more memorable to us than watching Big Brother on a large silver screen.

So why are there no films about a future where everything's great? An Earth where the air is clean and people live normal lives, and you can't buy eyeballs like you're buying Chicken McNuggets? And, of course, why don't people read books in any of these movies?

I think it's because books – even in our present day – are seen as signs of our past. Books represent an old fashioned way of doing things, signs of a former life and era and time, and futurists know this. So a sci-fi film would never have a scene with someone reading a book, any more than they'd have a scene where a person sits in their sleek sleep-pod and listens to Elvis Presley on a turntable. Or, if they did, the point would be that the person – like Captain Picard reading a book on the TV show *Star Trek: The Next Generation* – is fingering an *antique*.

True, both science fiction writers and film directors go for a reaction and an impact, and dystopia gives them all kinds of great angles to exploit. After all, if the future were too much like today, no one would pay $12 to see it; they'd just look out their windows instead. But also, most great science fiction or visions of the future are really criticisms of the present, and are not predictions. *Nineteen Eighty-Four* featured Big Brother not because Orwell necessarily though that it was going to happen (at least not in a literal way), but rather because he was trying to sound alarm bells so that Big Brother *didn't* happen – at the time, Big Brother already existed in small doses in communist countries (most notably in Stalin's Russia).

Orwell was not trying to guess tomorrow's lottery numbers. The same with Bradbury's *Fahrenheit 451*; he didn't think that firemen would one day really start fires instead of put them out. But instead, through his novel, he was trying to warn against a society that – because of cultural apathy – makes something obsolete to the point where it ceases to not only be irrelevant but is also thrown away.

In all these examples, and every one of these films or books, there really is something to their ideas of the future. Because even if they're getting the future *wrong* (and let's certainly hope most of them are), what each of them shows is that times change, life changes, and culture changes in real and profound ways, and the life that future generations lead will be much different from ours right now. If this seems unlikely, stop a ten-year-old and tell him what life was like when you were his age. And then watch in wonderment as he stares at you like you're a caveman. 'You didn't have cell phones? You didn't have the *Internet*?' And some of the things we would tell them that we *did* have (like rotary phones or even answering machines with cassette tapes) he wouldn't even know about, since those items have already been erased.

Or do the opposite. Find someone fifty years older than you are and listen to them tell you about when the fastest way to communicate was by telegram, or when radio was the biggest form (if not the only form) of entertainment you could have in your house. See if their world sounds anything like yours.

These are small examples, and somewhat silly ones, but put together I think they form a large testament to just how much time and culture and things can change. Even things that we thought, perhaps, would always stay the same; everything, simply everything, has the capacity for change.

The biggest change in the past fifty years, in terms of life on Earth, has been the introduction of the Internet and the abundance of gadgets that have arrived along with it: iPods, laptops, Blackberries, PDAs, eBook devices, not to mention cell phones, video cameras and portable video games. In fact, the science fiction movie *Terminator 3* was subtitled *The Rise of the Machines*, which could very well describe the first couple of years of this century.

The cumulative effect of all of these machines and inventions has been to truly transform the way we live our lives on almost every level. In fact, sometimes – with the ability to harness the world's collective intelligence and information, not to mention communicate with anyone anywhere in the blink of an eye – our modern times can seem a little bit like, well, sci-fi. Whether or not you think this is utopia or dystopia depends on your viewpoint, but one cannot possibly argue that things are staying the same.

'This is not, of course, the first such shift in our long history,' wrote Sven Birkerts in his 2006 book *The Gutenberg Elegies: The Fate of Reading in an Electronic Age*:

> In Greece, in the time of Socrates, several centuries after Homer, the dominant oral culture was overtaken by the writing technology. And in Europe another epochal transition was effected in the late fifteenth century after Gutenberg invented movable type. In both cases the longterm societal effects were overwhelming, as they will be for us in the years to come.

The societal effects that Birkerts describes are still very much with us, even a decade after the dawn of the Internet. Ripple effects are felt almost everywhere. And with every day, month and year that passes, new generations get more and more used to the idea of a constantly online life. This will play a part in the way that Digital Natives interact with nearly everyone: family, friends, teachers, lovers, employers.

This will also have an effect on the way they interact with forms of entertainment, such as music, movies, television and books. And *interact* is the key word, since future generations of Digital Natives won't be content to just *watch* or *listen* or *read*. Instead they'll want to participate in and interact somehow with the material they choose to consume and absorb, even if that just means consuming or absorbing the material whenever and however they want.

Because of this change in consumer habits – and given the need for Generation Upload to play an active role in its entertainment – books will have to evolve in the same way that music and movies have had to evolve. And it won't be the first time the written word has seen massive change. As Anna Quindlen described the process in her 1998 book *How Reading Changed My Life*, 'The clay tablet gave way to the scroll and then to the codex, the folded sheets that prefigured the book we hold and sell and treasure today.'

However, no matter how much we treasure the book, what's really important is the culture of ideas and innovation that books represent. It's this culture that's at stake, not the publishing companies or the fate of bookstores, or even the book itself as a physical form. That's all a sideshow to the main event. Whether or not the clay tablet evolves into the tablet PC, what should be at the heart of the conversation is a notion of literary culture and the idea that words can change the way we look at life.

'I would suggest that for all the passion and affection I bring to books, I have very little business caring for the future of the book,' wrote James J. O'Donnell in the 1996 essay collection *The Future of the Book*. 'Books are only secondary bearers of culture.'

Publishing itself is also ripe for a change, and with these new applications for reading we have to be prepared that they might not look anything like the pages in books we now know. It's important to remember that pages were invented to hold words; words were not invented to fill pages. While concrete poets may have used words in a certain order to portray the things they were writing about or to achieve a sublime overall effect, for the most part authors compose their works in words on a page simply because it's a means of communication.

Printed literature has always been a facsimile of what the author originally created, and the existence of dozens of classic books in many editions at once (ranging from fancy hard-backs to cheaply produced mass market paperbacks), proves that what we truly care about are the words, the story, the lives and characters that exist inside something like *Wuthering Heights*. Indeed, no one bristles at the thought that Heming-way handwrote his stories in longhand while standing up, and yet we read them printed on a page while sitting down. What matters least of all are the pages. Books are really just a deliv-ery option, and in the future consumers will simply have multiple options.

As I mentioned in Chapter 1, Booth Tarkington documented similar changes in his 1918 novel *The Magnificent Ambersons*, writing about the rise of the automobile and the encroaching cement tide of the city:

> Our electric extensions of ourselves simply by-pass space and time, and create problems of human involvement and organization for which there is no precedent. We may yet yearn for the simple days of the automobile and the superhighway.

And we may yet yearn for the simple days of a paperback book tucked into a backpack. The same way that some people yearn for a world where men always wore hats and cell phones weren't invented (not to mention phones them-selves), and big band was the big sound on the big radios found in the corner of most living rooms. But while nostalgia might be good for a pastime, it's suicide as a business model.

That culture and society move in such ways cannot be in doubt. All it takes is to see old movies to realize that we live in a much different world now than existed in the past. Take some-thing as simple as a telegram: sixty years ago it was the fastest way to send a message – a brief flurry of informative words – across the country or even entire continents. Telegrams battled time and space, transmitting pulses that somehow turned into words at the other end of the wire, hurtling across thousand of miles in seconds. Email now does the same thing, except it's even faster and much more efficient since, instead of traveling on poles stuck across the county like whiskers on a giant face, it involves ultra-thin fiber optic cables hiding underneath the ground (if not satellites spinning high above us in space). And yet despite all the technological change that occurred in the time between telegrams and email, the point of each is reassuringly the same: communication.

The segue from reading words in print to reading them on a computer screen will be a similar change in that the technology will simply be an aid in satisfying people's needs to immerse themselves in stories. Remember that books were really the second iteration of storytelling, the first being oral recitation. At the dawn of publishing, a printed copy of *The Iliad* would have seemed as silly as an electronic copy of *Pride and Prejudice* seems today.

Given everything we know, and everything we've been able to witness during the decades that have brought us the Internet revolution – a dozen tumultuous years that nobody could have predicted – all of these new inventions and ways of living will undoubtedly impact reading and publishing. Indeed, they already have; witness the massive layoffs in newspapers and magazines that can be directly attributed to the Web, not to mention the overall decline in reading and book sales. It would be foolhardy, if not terribly dangerous, not to realize this and see the connection. It's simply not possible that the

Internet is going to have an effect on every area of our lives *except* reading books. It has already had profound effects on the way people buy, write, produce and talk about books. So why not the books themselves?

In Hermann Hesse's breakthrough 1929 novel *Steppenwolf*, Mozart appears in the closing pages as part of an elaborate fantasy. Mozart presents Harry, the narrator, with an old radio. He then dials in some classical music, a concerto by Handel.

Harry is, at first, aghast; what is this miserable little device doing squeaking out the ethereal beauty of Handel? He almost can't stand to listen; it's close to torture. Slowly, however, he manages to hear the tune over the crackle and distortion; 'behind the slime and the croaking there was, sure enough, like an old master beneath a layer of dirt, the noble outline of that divine music.'

Mozart only laughs at Harry, letting the radio continue to play, allowing 'the murdered and murderous music' to 'ooze out and on.' Finally, still laughing, he upbraids Harry, telling him to get his mind off the mechanism and to concentrate on the music:

> Just listen, you poor creature, listen without either pathos or mockery, while far away behind the veil of this hopelessly idiotic and ridiculous apparatus the form of this divine music passes by. Pay attention and you will learn something. Observe how this crazy funnel apparently does the most stupid, the most useless and the most damnable thing in the world. It takes hold of some music played where you please, without distinction, stupid and coarse, lamentably distorted, to boot, and chucks it into space to land where it has no business to be; and yet after all this it cannot destroy the original spirit of the music; it can only demonstrate its own senseless mechanism, its

inane meddling and marring. Listen, then, you poor thing. Listen well.

With the rise in digital reading, computers, laptops, and cell phones – the same as the radio in the above passage – will merely be apparatuses, the 'crazy funnels' into which we will pour the great words of the past, present and future. True, they won't be the same as a printed book, but that doesn't mean they will 'destroy the original spirit' of the books we knew and loved. In the same way that Handel coming out of a tinny speaker still possesses the genius of Handel, Fitzgerald's gorgeous prose will continue to be gorgeous even when rendered on the screen of a computer.

However, there are plenty of people who feel – like Hesse's narrator feels about music – that words have 'no business to be' anywhere other than on the printed page. They feel that reading on any kind of electronic device is blasphemy, and that books are sacred. And yet, considering that the alternative is silence – that if upcoming generations don't read digitally there's a good chance they won't read at all – then through whatever mechanism it takes to get words in front of a pair of curious human eyes, or wherever those words end up, the important thing is that they are *read*.

notes

Introduction

Sven Birkerts (2006) *The Gutenberg Elegies: The Fate of Reading in an Electronic Age*. Faber & Faber, London.

Stop the presses

Chapter 1: Byte flyte

George P. Landlow (1996) Twenty minutes into the future, or how are we moving beyond the book. In: *The Future of the Book* (ed. Geoffrey Nunberg). University of California Press.

Chris Anderson (2006) *The Long Tail: Why the Future of Business Is Selling More of Less*. Hyperion, New York.

Anthony Burgess (1984) *99 Novels*. Summit Books, New York.

The London Book Fair (2007) Digitise or die: what is the future of the book? The London Book Fair Website, http://www.londonbookfair.co.uk/page.cfm/Link=198/t=m/goSection=12.

Mike Elgan (2007) Why e-books are bound to fail. *Computerworld*, 27 April.

Andrew Marr (2007) Curling up with a good ebook. *The Guardian*, 11 May.

John Lanchester (2007) It's a steal. *The Guardian*, 7 April.

Cory Doctorow (2007) You *do* like reading off a computer screen. *Locus*, March.

The Economist (2007) Not bound by anything. *The Economist*, 22 March.

Booth Tarkington (1998) *The Magnificent Ambersons*. Modern Library Edition.

Chapter 2: Us and them

C. P. Snow (1998) *The Two Cultures*. Cambridge University Press, Cambridge.

Bob Thompson (2006) Explosive words. *The Washington Post*, 22 May.

E. Annie Proulx (1994) Books on top. *The New York Times*, 24 May.

Ray Bradbury (1991) *Fahrenheit 451*. Del Rey, New York.

National Endowment for the Arts (2004) *Reading at Risk: a Survey of Literary Reading in America*.

Pat Walsh (2005) *78 Reasons Why Your Book May Never Be Published and 14 Reasons Why it Just Might*. Penguin, New York.

Douglas Rushkoff (1999) *Playing the Future: What We Can Learn from Digital Kids*. Riverhead Books, New York.

Pip Coburn (2006) *The Change Function: Why Some Technologies Take Off and Others Crash and Burn*. Penguin Books, New York.

Steven Levy (2006) *The Perfect Thing: How the iPod Shuffles Commerce, Culture, and Coolness*. Simon and Schuster, New York.

Jon Pareles (2002) David Bowie, 21st-century entrepreneur. *The New York Times*, 9 June.

Jacob Weisberg (2000) The good e-book. *The New York Times*, 4 June.

Marshall McLuhan (1962) *The Gutenberg Galaxy: The Making of Typographic Man*. University of Toronto Press.

Motoko Rich (2006) Digital publishing is scrambling the industry's rules. *The New York Times*, 5 June.

Nicholas A. Basbanes (2005) *Every Book Its Reader: The Power of the Printed Word to Stir the Soul*. HarperCollins, New York.

Chapter 3: Newspapers are no longer news

Nicholas Negroponte (1996) The DNA of Information. *Being Digital*. Vintage, New York.

Anna Quindlen (1998) *How Books Changed My Life*. The Library of Contemporary Thought/Ballantine Books.

Reuters (2006) Newspaper website readership up 31%. 4 October.

Michael Kinsley (2006) Do newspapers have a future? *Time*, 25 September.

Elizabeth M. Neiva (1995) Chain building: the consolidation of the American newspaper industry, 1995–1980. *Business and Economic History*, **24**(1).

C. P. Snow (1998) *The Two Cultures*. Cambridge University Press, Cambridge.

Steven Levy (2006) *The Perfect Thing: How the iPod Shuffles Commerce, Culture, and Coolness*. Simon and Schuster, New York.

John Freeman (2007) NBCC campaign to save book reviews. *Critical Mass*, 23 April, http://bookcriticscircle.blogspot.com/2007/04/nbcc-will-fight-these-cut-backs.html.

Art Winslow (2007) The new book burning. *The Huffington Post*, 25 April, http://www.huffingtonpost.com/art-winslow/the-new-book-burning_b_46820.html.

Michael Connolly (2007) The folly of downsizing book reviews. *The Los Angeles Times*, 29 April.

Pat Holt (2007) Book critics: are we driving readers away? *Holt Uncensored*, 30 April, http://www.holtuncensored.com/members/index.html.

Motoko Rich (2007) Are book reviewers out of print? *The New York Times*, 2 May.

Totally wired

Chapter 4: Generation download

Joseph Menn (2003) *All the Rave: The Rise and Fall of Shawn Fanning's Napster*. Crown Business, New York.

Drew Dernavich (2006) Sorry, I think I just pressed shuffle. *The New Yorker*, 6 January.

David Kusek and Gerd Leonhard (2005) *The Future of Music: Manifesto for the Digital Music Revolution*. Berklee Press, Boston, MA.

Neil Howe and William Strauss (2000) *Millennials Rising: The Next Great Generation*. Vintage, New York.

Alex Williams (2006) The graying of the record store. *The New York Times*, 16 July.

Associated Press (2006) Tower Records victim of iPod era. 10 October.

MarketWatch (2006) EMI Music CEO says the CD is 'dead'. MarketWatch, 27 October.

Claudia H. Deutsch (2006) A milestore for iTunes; a windfall for a downloader. *The New York Times*, 24 February.

Matt Richtel (2007) It don't mean a thing if you ain't got that ping. *The New York Times*, 22 April.

Sven Birkerts (2006) *The Gutenberg Elegies: The Fate of Reading in an Electronic Age*. Faber & Faber, London.

Marshall McLuhan (1964) *Understanding Media: The Extensions of Man*. New American Library, New York.

Chapter 5: Generation upload

Marshall McLuhan (1962) *The Gutenberg Galaxy: The Making of Typographic Man*. University of Toronto Press.

Tu Thanh Ha (2006) 'Star Wars Kid' cuts a deal with his tormentors. *Toronto Globe and Mail*, 4 July.

Bob Garfield (2006) YouTube vs. boob tube. *Wired*, December.

Virginia Heffernan (2006) Web guitar wizard revealed at last. *The New York Times*, 27 August.

Matthew Klam (2006) The online auteurs. *The New York Times*, 12 November.

Joshua Davis (2006) The secret world of lonelygirl. *Wired*, December.

Reuters (2006) New rock stars use Web videos to win fans. 3 September.

Clive Thompson (2007) Sex, drugs and updating your blog. *The New York Times*, 13 May.

Eric Steuer (2006) (Beastie) Boys on film. *Wired*, April.

Eric Steuer (2006) The infinite album. *Wired*, September.

Chris M. Walsh (2006) New beck album denied U.K. chart eligibility. *Billboard*, 4 October.

Don Tapscott (1998) *Growing Up Digital: The Rise of the Net Generation*. McGraw-Hill, New York.

Emily Nussbaum (2007) Say anything. *New York Magazine*, 12 February.

Anna Quindlen (1998) *How Reading Changed My Life*. The Library of Contemporary Thought, Ballantine, New York.

Steven Levy (2006) *The Perfect Thing: How the iPod Shuffles Commerce, Culture, and Coolness*. Simon and Schuster, New York.

Douglas Rushkoff (1999) *Playing the Future: What We Can Learn from Digital Kids*. Riverhead Books, New York.

Chapter 6: On demand everything

Johnnie L. Roberts (2006) Why prime time's now your time. *Newsweek*, 30 October.

Jeff Jarvis (2006) Opinion. *The Guardian*, 2 October.

Don Tapscott (1998) *Growing Up Digital: The Rise of the Net Generation*. McGraw-Hill, New York.

Sean Smith (2006) When the 'bubble' bursts. *Newsweek*, 23 January.

Paul J. Gough (2006) Spielberg calls for responsible TV. *The Hollywood Reporter*. 20 November.

Xeni Jardin (2005) Thinking outside the box office. *Wired*, December.

Thomas H. Davenport and John C. Beck (2001) *The Attention Economy: Understanding the New Currency of Business*. Harvard Business School Press, Harvard, MA.

Ira Boudway (2007) The media diaries. *New York Magazine*, 15 January.

Jason Epstein (2006) Books@Google. *The New York Review of Books*, 19 October.

Chapter 7: eBooks and the revolution that didn't happen

Walt Crawford (2006) Why aren't ebooks more successful? *Econtent*, October.

David D. Kirkpatrick (2001) Forecasts of an e-book era were, it seems, premature. *The New York Times*, 28 August.

Charles McGrath (2006) Can't judge an e-book by its screen? Well, maybe you can. *The New York Times*, 24 November.

Pip Coburn (2006) *The Change Function: Why Some Technologies Take Off and Others Crash and Burn*. Penguin, New York.

Steven Levy (2006) *The Perfect Thing: How the iPod Shuffles Commerce, Culture, and Coolness*. Simon and Schuster, New York.

Associated Press (2007) Sales of albums plunge, but digital downloads soar. 5 January.

Steve Jobs (2007) Thoughts on music. Apple Website, 6 February, http://www.apple.com/hotnews/thoughtsonmusic/.

Thomas Crapton (2007) EMI dropping copy limits on online music. *The New York Times*, 3 April.

Blake Wilson (2006) Has the iPod for books arrived? *Slate*, 13 October, http://www.slate.com/id/2151525/.

Paul Duguid (1996) Material matters: the past and futurology of the book. In: *The Future of the Book* (ed. Geoffrey Nunberg). University of California Press.

Saying goodbye to the book

Chapter 8: Writers in a digital future

George Gissing (1980) *New Grub Street*. Penguin English Library, New York.

Marshall McLuhan (1964) *Understanding Media: The Extensions of Man*. New American Library, New York.

Anna Quindlen (1998) *How Reading Changed My Life*. The Library of Contemporary Thought, Ballantine, New York.

Jay David Bolter (2001) *Writing Space: Computers, Hypertext and the Remediation of Print*. Lawrence Erlbaum Associates, New York.

Robert Coover (1992) The end of books. *The New York Times*. 21 June.

Luca Toschi, (1996) Hypertext and authorship. In: *The Future of the Book* (ed. Geoffrey Nunberg). University of California Press.

Norman Mailer (1994) *Advertisements for Myself*. Flamingo Modern Classics, New York.

Pat Walsh (2005) *78 Reasons Why Your Book May Never Be Published and 14 Reasons Why it Just Might*. Penguin, New York.

Cory Doctorow (2006) Giving it away. *Forbes*, 1 December.

Milan Kundera (2007) *The Curtain: An Essay in Seven Parts*. HarperCollins, New York.

Chapter 9: Readers in a digital future

Anna Quindlen (1998) *How Reading Changed My Life*. The Library of Contemporary Thought, Ballantine, New York.

Chip Walter (2005) Kryder's Law. *Scientific American*, August.

Charles McGrath (2006) Can't judge an e-book by its screen? Well, maybe you can. *The New York Times*, 24 November.

Marshall McLuhan (1962) *The Gutenberg Galaxy: The Making of Typographic Man*. University of Toronto Press.

W. Somerset Maugham (1963) *Collected Short Stories*, Vol. 4. Penguin, London.

David Kusek and Gerd Leonhard (2005) *The Future of Music: Manifesto for the Digital Music Revolution*. Berklee Press, Boston, MA.

Anthony Burgess (1984) *99 Novels*. Summit Books, New York.

Jacob Weisberg (2000) The good e-book. *The New York Times*, 4 June.

Maureen Corrigan (2005) *Leave Me Alone, I'm Reading: Finding and Losing Myself in Books*. Random House, New York.

Edward Tenner (2007) The prestigious inconvenience of print. *The Chronicle of Higher Education*, 9 March.

Burt Helm (2005) Curling up with a good e-book. *Business Week Online*, 29 January, http://www.businessweek.com/technology/content/dec2005/tc20051229_155542.htm.

Cory Doctorow (2006) Giving it away. *Forbes*, 1 December.

Chapter 10: Will books disappear?

Al Ries and Laura Ries (2004) *The Fall of Advertising and the Rise of PR*. HarperCollins, New York.

Julie Bosman (2007) A Princeton maverick succumbs to a cultural shift. *The New York Times*, 3 January.

Mike Peed (2007) Booksellers fight back as 5 new stores open. *The New York Sun*, 15 January.

Associated Press (2006) Indie bookstores tackle internet. 8 October.

Dan Barry (2007) A requiem for reading in a smoldering pyre of books. *The New York Times*, 3 June.

David Kirkpatrick (2000) Stephen King sows dread in publishers with his latest e-tale. *The New York Times*, 24 July.

Jon Katz (2000) Stephen King's net horror story. Slashdot, 4 December, http://slashdot.org/features/00/11/30/1238204.shtml.

Seth Godin (2006) *Advice for Authors*. Sethgodin.com, 2 August. http://sethgodin.typepad.com/seths_blog/2006/08/advice_for_auth.html

Elaura Niles (2005) *Some Writers Deserve to Starve!: 31 Brutal Truths About the Publishing Industry*. Writer's Digest Books, Cincinnatti, OH.

Chris Anderson (2006) *The Long Tail: Why the Future of Business Is Selling More of Less*. Hyperion, New York.

Emily Maltby (2006) An ATM for books. *Fortune Small Business Magazine*, 14 December.

Marshall McLuhan (1964) *Understanding Media: The Extensions of Man*. New American Library, New York.

Sean Wilsey (2006) Why John Updike is so wrong about digitized books. *Time*, 31 May.

Zadie Smith (2007) Fail better. *The Guardian*, 13 January.

Afterword

Sven Birkerts (2006) *The Gutenberg Elegies: The Fate of Reading in an Electronic Age*. Faber & Faber, London.

Anna Quindlen (1998) *How Reading Changed My Life*. The Library of Contemporary Thought, Ballantine, New York.

James J. O'Donnell (1996) Trithemius, McLuhan, Cassiodorus. In: *The Future of the Book* (ed. Geoffrey Nunberg). University of California Press.

Herman Hesse (2002) *Steppenwolf*. Picador USA, New York.

acknowledgements

I would like to express my extreme thanks to Sara Abdulla, whose idea it was for me to write this book. Thanks also to Alexandra Dawe for shepherding my manuscript through its final stages. I would also like to acknowledge the talented people I have had the honor to work with at Holtzbrinck, in particular John Sargent, Brian Napack and Fritz Foy. In addition, I'd like to thank a number of other people within the Macmillan family who have shown support and encouragement over the years, including Alison Lazarus, Steve Cohen, Peter Garabedian, Richard Charkin and Stefan von Holtzbrinck.

The Tom Stoppard quote on p. viii is an excerpt from an interview by Daphne Merkin, 'Playing With Ideas' from *The New York Times Magazine*, 26 November 2006. Copyright © 2006 by The New York Times Company. Reprinted by permission.

Index

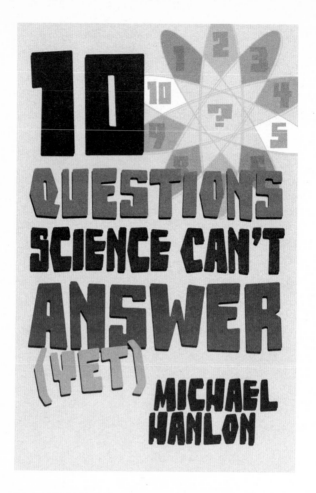

10 QUESTIONS SCIENCE CAN'T ANSWER (YET)
by Michael Hanlon
MACMILLAN; ISBN: 978–0–230–51758–5; £14.99/$24.95;
HARDCOVER

Ten examples of the different kinds of mysteries scientists have
yet to crack, some plausible theories and a few less likely ones.
Find out why these questions, and others like them, remain,
and when – or indeed whether – we might get some answers.

order now from www.macmillanscience.com

THE ADVOCATE OF INDUSTRY AND ENTERPRISE, AND JOURNAL OF MECHANICAL AND OTHER IMPROVEMENTS.

VOLUME 1   NEW-YORK, THURSDAY, AUGUST 28, 1845.   NUMBER 1.

Bringing science & technology alive
for 160 years

Whether reporting on a radical ship design in 1854, an arctic exploration gone wrong in 1904, Loop Quantum Gravity, Time Before the Big Bang, or how a Mars robot works, SCIENTIFIC AMERICAN has always set the standard for intelligent discussion of science and technology. Today, there is simply no other publication that commands the same respect of leading scientists, while inspiring students and readers.

SCIENTIFIC AMERICAN

where the best minds come for the best science journalism

www.sciam.com

A photocopy of this order form is acceptable.

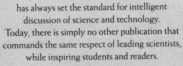

P4MS320

SCIENTIFIC AMERICAN **_Yes!_** Send me 12 issues – a full year – of **SCIENTIFIC AMERICAN** for:
❏ $34.97 [U.S. addresses only]   ❏ $44.00 Elsewhere

Name: _____

Address: _____

City: _____

State/Country: _____   Zip/Postal Code: _____

❏ Payment enclosed.   ❏ Bill me later.   Charge my: ❏ MasterCard ❏ Visa ❏ American Express

Card #: _____   Exp. Date: _____

Signature: _____

MAIL ORDER TO: SCIENTIFIC AMERICAN, PO BOX 3186, HARLAN, IA 51593-0377 USA

Remit in: U.S. dollars drawn on a U.S. bank, Pound Sterling checks drawn on a U.K. bank, Euro Checks drawn on a EU bank. Current exchange rates are acceptable. Price includes postage. GST is included; please add appropriate PST. BN# 127387652RT. QST# Q1015332537 in Canada. 2005

nature

The leading international weekly journal of science

An illustrious tradition of editorial excellence.

Since 1869, *Nature* has consistently published the best of scientific journalism and primary research, including:

1896: X-rays discovered
1939: Nuclear fission explained
1952: Structure of DNA discovered
1974: CFCs found destroying stratospheric ozone
1977: The obese gene cloned
1995: Discovery of extra-solar planets
1997: Dolly the sheep born
2001: Human genome sequenced

Subscribe to *Nature* today and get 51 issues of high quality science with full online access, including FREE access to news@nature.com, *the* authoritative online science news service.

Visit: www.nature.com/nature/macscioffer and secure a **20% discount.**

nature publishing group npg